BEYOND THE SELF

BEYOND THE SELF

Conversations between Buddhism and Neuroscience

MATTHIEU RICARD AND WOLF SINGER

The MIT Press
Cambridge, Massachusetts
London, England

First MIT Press paperback edition, 2018
© 2017 Allary Editions

Published by special arrangement with Allary Editions in conjunction with their duly appointed agent 2 Seas Literary Agency.

This book was set in Scala by Toppan Best-set Premedia Limited. Printed and bound in the United States of America.

Library of Congress Cataloging-in-Publication Data

Names: Ricard, Matthieu, author.
Title: Beyond the self : conversations between Buddhism and neuroscience / Matthieu Ricard and Wolf Singer.
Other titles: Cerveau & méditation. English
Description: Cambridge, MA : MIT Press, 2017. | Includes bibliographical references and index.
Identifiers: LCCN 2017010026 | ISBN 9780262036948 (hardcover : alk. paper); 9780262536141 (paperback)
Subjects: LCSH: Neurosciences--Religious aspects--Buddhism. | Buddhism--Psychology.
Classification: LCC BQ4570.N48 R5313 2017 | DDC 294.3/3153--dc23 LC record available at https://lccn.loc.gov/2017010026

10 9 8 7 6 5 4

CONTENTS

PREFACE

It all started in London in 2005, when we first discussed the theme of consciousness. That same year we saw each other in Washington, DC, to talk about the neuronal basis of meditation at a meeting organized by the Mind and Life Institute.[1] For eight years, we took every chance we could to continue our exchanges all over the world, twice in Nepal, in the rainforests of Thailand, and with His Holiness the Dalai Lama in Dharamsala, India.[2] This book is the result of this extended conversation, nourished by friendship and our shared interests.

The dialogue between Western science and Buddhism stands out from the often difficult debate between science and religion. It is true that Buddhism is not a religion in the sense we usually understand in the West. It is not based on the notion of a creator and therefore does not require an act of faith. Buddhism could be defined as a "science of the mind" and a path of transformation that leads from confusion to wisdom, from suffering to freedom. It shares with the sciences the ability to examine the mind empirically. This is what makes the dialogue between a Buddhist monk and a neuroscientist possible and fruitful: a broad range of questions can be broached, from quantum physics to ethical matters.

We have attempted to compare the Western and Eastern perspectives, the different theories concerning the constitution of the self and the nature of consciousness as seen by the scientific and contemplative points of view. Until recently, most Western philosophies have been built around the separation of mind and matter. Scientific theories that are today attempting to explain how the brain works bear the mark of this dualism. Buddhism, meanwhile, has proposed a nondualistic approach to reality from the start. The cognitive sciences see consciousness as being inscribed in the body, society, and culture.

Hundreds of books and articles have been dedicated to theories of knowledge, meditation, the idea of the self, emotions, the existence of free will, and the nature of consciousness. Our aim here is not to make an inventory of the many points of view that exist on these subjects. Rather, our objective is to confront two perspectives anchored in rich traditions: the contemplative Buddhist practice, and epistemology and research in neuroscience. We were able to bring together our experiences and skills to try and answer the following questions: Are the various states of consciousness arrived at through meditation and training the mind linked to neuronal processes? If so, in what way does the correlation operate?

This dialogue is only a modest contribution to an immense field confronting the points of view and knowledge about the brain and consciousness of scientists and people who meditate—in other words, the meeting between first- and third-person knowledge. The lines that follow take this path, and we feel humility in front of the size of the task. We sometimes allow ourselves to be swept away by the themes close to our hearts, which translate in certain places into changes in direction or repetitions. We made the choice to retain the authenticity of the dialogue because it is rare and productive to develop an exchange over such a long period. We would nevertheless like to apologize to our readers for what may seem like an oversight.

This dialogue allowed us to make progress in our mutual understanding of the themes we addressed. By inviting our readers to join us, we hope they too will benefit from our years of work and investigation into the fundamental aspects of human life.

MEDITATION AND THE BRAIN

A SCIENCE OF MIND

Our capacity to learn is far superior to that of other animals. Can we, with training, develop our mental skills, as we do for our physical skills? Can training the mind make us more attentive, altruistic, and serene? These questions have been explored for 20 years by neuroscientists and psychologists who collaborate with people who meditate. Can we learn to manage our disturbing emotions in an optimal way? What are the functional and structural transformations that occur in the brain due to different types of meditation? How much time is needed to observe transformations like this in people new to meditation?

Matthieu: Although one finds in the Buddhist literature many treatises on "traditional sciences"—medicine, cosmology, botanic, logic, and so on—Tibetan Buddhism has not endeavored to the same extent as Western civilizations to expand its knowledge of the world through the natural sciences. Rather it has pursued an exhaustive investigation of the mind for 2,500 years and has accumulated, in an empirical way, a wealth of experiential findings over the centuries. A great number of people have dedicated their whole lives to this contemplative science. Modern Western psychology began with William James just over a century ago. I can't help remembering the remark made by Stephen Kosslyn, then chair of the psychology department at Harvard, at the Mind and Life meeting on "Investigating the Mind," which took place at MIT in 2003. He started his presentation by saying, "I want to begin with a declaration of humility in the face of the sheer amount of data that the contemplatives are bringing to modern psychology."

It does not suffice to ponder how the human psyche works and elaborate complex theories about it, as, for instance, Freud did. Such intellectual constructs cannot replace two millennia of direct investigation of the workings of mind through penetrating introspection conducted with trained minds that have become both stable and clear. Any sophisticated theory that came out of a brilliant mind but does not rest on empirical evidence cannot be compared with the cumulated experience of hundreds of people who have each a good part of their lives fathomed the subtlest aspects of mind through direct experience. Using empirical approaches undertaken with the right instrument of a well-trained mind, these contemplatives have found efficient ways to achieve a gradual transformation of emotions, moods, and traits, and to erode even the most entrenched tendencies that are detrimental to an optimal way of being. Such achievements can change the quality of every moment of our lives through enhancing fundamental human characteristics such as lovingkindness, inner freedom, inner peace, and inner strength.

Wolf: Can you be more specific with this rather bold claim? Why should what nature gave us be fundamentally negative, requiring special mental practice for its elimination, and why should this approach be superior to conventional education or, if conflicts arise, to psychotherapy in its various forms, including psychoanalysis?

Matthieu: What nature gave us is by no means entirely negative; it is just a baseline. Most of our innate capacities remain dormant unless we do something, through training, for instance, to bring them to an optimal, functional point. We all know that our mind can be our best friend or our worst enemy. The mind that nature gave us does have the potential for immense goodness, but it also creates a lot of unnecessary suffering for ourselves and others. If we take an honest look at ourselves, then we must acknowledge that we are a mixture of light and shadow, of good qualities and defects. Is this the best we can be? Is that an optimal way of being? These questions are worth asking, particularly if we consider that some kind of change is both desirable and possible.

Few people would honestly argue that there is nothing worth improving about the way they live and the way they experience the world. Some people regard their own particular weaknesses and conflicting emotions as a valuable and distinct part of their "personality," as something that contributes to the fullness of their lives. They believe that this is what makes them unique and argue that they should accept themselves as they are. But isn't this an easy way to giving up on the idea of improving the quality of their lives, which would cost only some reasoning and effort?

Our mind is often filled with troubles. We spend a great deal of time consumed by painful thoughts, anxiety, or anger. We often wish we could manage our emotions to the point where we could be free of the mental states that disturb and obscure the mind. It is easier indeed, in our confusion about how to achieve this kind of mastery, to adopt the view that this is all "normal," that this is "human nature." Everything found in nature is "natural," but that does not necessarily make it desirable. Disease, for example, comes to everybody and is perfectly natural, but does this prevent us from trying to cure it?

Nobody wakes up in the morning and thinks, "I wish I could suffer for the whole day and, if possible, for my whole life." Whatever we are occupied with, we always hope we will get some benefit or satisfaction out of it, either for ourselves or others, or at least a reduction of our suffering. If we thought nothing would come of our activities but misery, we wouldn't do anything at all, and we would fall into despair.

We don't find anything strange about spending years learning to walk, read and write, or acquire professional skills. We spend hours doing physical exercises to get our bodies into shape. Sometimes we expend tremendous physical energy pedaling a stationary bike that goes nowhere. To sustain such tasks requires at least some interest or enthusiasm. This interest comes from believing that these efforts are going to benefit us in the long run. Working with the mind follows the same logic. How could it be subject to change without any effort, just from wishing alone? We cannot learn to ski by practicing a few minutes once a year.

We spend a lot of effort improving the external conditions of our lives, but in the end it is always the mind that creates our experience of the world and translates this experience into either well-being or suffering. If we transform our way of perceiving things, then we transform the quality of our lives. This kind of transformation is brought about by the form of mind training known as meditation.

We significantly underestimate our capacity for change. Our character traits remain the same as long as we do nothing to change them and as long as we continue to tolerate and reinforce our habits and patterns, thought after thought. The truth is that the state that we call "normal" is just a starting point and not the goal we should set for ourselves. Our life is worth much more than that. It is possible, little by little, to arrive at an optimal way of being.

Nature also gave us the possibility to understand our potential for change, no matter who we are now and what we have done. This notion is a powerful source of inspiration for engaging in a process of inner transformation. You may not succeed easily, but at least be encouraged by such an idea; you can put all your energy into such a transformation, which is already in itself a healing process.

Modern conventional education does not focus on transforming the mind and cultivating basic human qualities such as lovingkindness and mindfulness. As we will see later, Buddhist contemplative science has many things in common with cognitive therapies, in particular with those using mindfulness as a foundation for remedying mental imbalance. As for psychoanalysis, it seems to encourage rumination and explore endlessly the details and intricacies of the clouds of mental confusion and self-centeredness that mask the most fundamental aspect of mind: luminous awareness.

Wolf: So rumination would be the opposite of what you do during meditation?

Matthieu: Totally opposite. It is also well known that constant rumination is one of the main symptoms of depression.

Wolf: It is encouraging for our dialogue to have contrasting views on strategies to cure the mind. I suspect that the practice of meditation is often misunderstood. I have had little practice with it, but I learned to see what it is not: it is not an attempt to confront oneself with unresolved problems to search for their causes and eliminate them. It appears to be quite the contrary.

Matthieu: When one looks at the process of rumination, it is easy to see what a troublemaker it is. What we need is to gain freedom from the mental chain reactions that rumination endlessly perpetuates. One should learn to let thoughts arise and be freed to go as soon as they arise, instead of letting them invade one's mind. In the freshness of the present moment, the past is gone, the future is not yet born, and if one remains in pure mindfulness and freedom, potentially disturbing thoughts arise and go without leaving a trace.

Wolf: You have said in one of your books that every human being possesses in his mind a "nugget of gold," a kernel of purity and positive qualities that is, however, concealed and overshadowed by a host of negative traits and emotions that deform his perceptions and are the major cause of suffering. To me, this sounds like an overly optimistic and untested hypothesis. It sounds like Rousseau's dreams and seems to be contradicted by cases like that of the feral child Kaspar Hauser. We are what evolution imprinted by genes and culture via education, moral norms, and social conventions. What then is the "golden nugget"?

Matthieu: A piece of gold that remains deeply buried in its ore, in a rock, or in the mud. The gold does not lose its intrinsic purity, but its value is not actualized. Likewise, to be fully expressed, our human potential needs to meet with suitable conditions.

AWARENESS AND MENTAL CONSTRUCTS

Matthieu: The idea of an unspoiled basic nature of consciousness is not a naïve assessment of human nature. It is based on reasoning and introspective experience. If we consider thoughts, emotions, feelings, and any other mental events, they all have a common denominator, which is the capacity of knowing. In Buddhism, this basic quality of consciousness is called the *fundamental luminous nature of mind*. It is luminous in the sense that it throws light on the outer world through our perceptions and on our inner world through our feelings, thoughts, memories of the past, anticipation of the future, and awareness of the present moment. It is luminous in contrast to an inanimate object, which is completely dark in terms of cognition.

Let's use this image of light. If you have a torch and you light up a beautiful smiling face or an angry face, a mountain of jewels or a heap of garbage, then the light does not become kind or angry, valuable or dirty. Another image is that of a mirror. What makes a mirror special is that it can reflect all kinds of images, but none of them belongs to, penetrates, or stays in the mirror. If they did, then all these images would superimpose, and the mirror would become useless. Likewise, the basic quality of the mind allows all mental constructs—love and anger, joy and jealousy, pleasure and pain—to arise but is not altered by them. Mental events do not belong intrinsically to the most fundamental aspect of consciousness. They simply occur within the space of awareness, of various moments of consciousness, and are made possible by this basic awareness. This quality can thus be called *basic cognition, pure awareness,* or *the most fundamental nature of mind.*

Wolf: What you said has two implications. One is that you seem to attribute value to stability or objectivity; it's like a validation criterion. The second is that you dissociate conscious awareness from its contents. You assume that a platform in the brain functions like an ideal mirror, not introducing any distortions by itself, not being influenced by the content it reflects. Are you defending a dualistic stance, a dichotomy between an immaculate mind and an observer, on the one hand, and the contents in this mind that are then fraught with all kinds of interferences and distortions? Contemporary views on the organization of the brain deny clear distinctions between sensory and executive functions and interpret consciousness as an emergent property of the integral functions of the brain. Thus, I have problems with the distinction between an immaculate mirror and reflected contents. I cannot conceive of an empty platform of consciousness—if it is empty, then it would just not exist; it would not be defined either.

Matthieu: Not at all. It is not a duality. There are not two streams of consciousness. It has more to do with various aspects of conscious-ness: a fundamental aspect, pure awareness, which is always there; and adventitious aspects, the mental constructs, which are always changing. We should rather speak of a continuity. Consciousness, at all levels, is but one dynamic flow made up of instants of awareness, with or without content. At any time behind the screen of thoughts, one can recognize a pure cognitive faculty that is the ground of all thoughts.

Wolf: This would then require at least two distinct entities: an empty space, which acts as a vessel with all the qualities you described; and the contents that, however muddled, do not affect the vessel.

Matthieu: Why two entities? The mind can be aware of itself without requiring a second mind to do so. One aspect of the mind, the most fundamental aspect of it, pure awareness, can also be awareness of itself without requiring a second observer. If the mirror and its contents bother you, then pure consciousness could also be compared to a piece of clay and mental constructs to the various shapes that the same clay can assume. No matter what shape you give to the clay, the clay is always there and never essentially changed.

Wolf: To have such an immaculate inner eye, such an ideal mirror that remains unaffected by and entirely decoupled from all emotions would—in my mind—requires a dissociation of the personality. There would be the immaculate observer, detached from emotions, affections, and misperceptions, and then there would be the other one, also part of you, who gets enmeshed in conflicts and misperceives situations because he has deeply fallen in love or is disappointed. Is the mental practice a tool to achieve such a dissociation of the self? What is your experience here? Is the creation of such dissociation—if this is what meditation aims at—not a hazardous experiment?

Matthieu: The point is not to fragment the self but to use the capacity of the mind to observe and know itself to free oneself from suffering. We actually speak of *nondual self-illuminating awareness*, which emphasizes this point. There is no need for a dissociation of personality because the mind has the inherent faculty to observe itself, just as a flame does not need a second flame to light itself up. Its own luminosity suffices.

The practical point of all this is that you can look at your thoughts, including strong emotions, from the perspective given by pure mindfulness. Thoughts are manifestations of pure awareness, just like waves that surge from and dissolve back into the ocean. The ocean and waves are not two intrinsically separate things. Usually, we are so taken by the content of thoughts that we fully identify ourselves with our thoughts and are unaware of the fundamental nature of consciousness, pure awareness. Because of that we are easily deluded, and we suffer.

The entire Buddhist path is about various ways to get rid of delusion. Take the example of a strong experience of malevolent anger. We become one with anger. Anger fills our whole mental landscape and projects its distortion of reality on people and events. When we become overwhelmed by anger, we cannot dissociate from it. We also perpetuate a vicious cycle of affliction by rekindling anger each time we see or remember the person who made us angry. Although anger is clearly not an enjoyable state of mind, we cannot help triggering it over and over again, like adding more and more wood to the fire. We thus become addicted to the cause of suffering. But if we dissociate from anger and look at it dispassionately with bare mindfulness, then we can see that it is just a bunch of thoughts and not something fearsome. Anger does not carry weapons, it does not burn like a fire or crush one like a rock; it is nothing more than a product of our mind.

Wolf: Doesn't it follow that positive emotions are equally detrimental because they lead to misperceptions and hence to suffering?

Matthieu: Not necessarily. It all depends on whether a mental event distorts reality. If the mind recognizes, for instance, that all beings aspire to be free from suffering and becomes filled with altruistic love and the strong wish to free them from suffering, then as long as it has this wisdom component, it can remain attuned to reality. It recognizes the interdependence of all beings, acknowledges their common wish to avoid suffering and achieve happiness, and ascertains the deep causes of their suffering. If, in addition to this, altruistic love is not biased by our various attachments and grasping, then it will not be afflictive. Instead of obscuring wisdom, it will manifest as the natural expression of wisdom.

But to conclude the analysis of anger, instead of "being" the anger and fully identifying with it, we must simply look at anger and keep our bare attention on it. When we do so, what happens? Just as when we cease to add wood to a fire, the fire soon dies out; anger cannot sustain itself for long under the gaze of mindfulness. It simply fades away.

Wolf: And so would love, empathy, sorrow, and all the other strong emotions. Do you aim for a clear mind without emotions? I doubt that such emotion-free human beings can survive and reproduce unless they have the privilege to live in a highly protected environment.

WORKING WITH EMOTIONS

Matthieu: The aim is not to cease to experience emotions but to avoid being enslaved by them. In Western languages, the word *emotion* comes from the Latin root *emove*, which means "to set in motion." An emotion is what sets the mind in motion, but much depends on how it does so. Your mind can be set in motion by the wish to alleviate someone's suffering. This is not afflictive. In addition, it does not make sense to try to block the arising of thoughts and emotions because they will surge in the mind anyway. The important point is what happens next. If afflictive emotions invade the mind, then you are in trouble. If, at the moment they arise, you find a way to let them undo themselves and vanish, then you have skillfully dealt with them.

By freeing anger, for instance, as it arises, we have avoided two unpractical ways of dealing with it. We did not let anger explode, with all the negative consequences that arise from such outbursts, such as hurting others, destroying our inner peace, and reinforcing our tendency to become angry often and easily. We also avoided merely suppressing anger, putting a lid on it while leaving it intact, like a time bomb, in some dark corner of our mind. We dealt with anger in an intelligent way, by letting its flames vanish. If we do so repeatedly, then anger will begin to arise less often and less strongly. Thus, the habitual tendency of becoming angry will gradually become eroded, and our traits will be transformed.

Wolf: What you have to learn then is to adopt a much more subtle approach to your internal emotional theater, to learn to identify with much higher resolution the various connotations of your feelings.

Matthieu: That's right. In the beginning, it is difficult to do it as soon as an emotion arises, but if you become increasingly familiar with such an approach, it becomes quite natural. Whenever anger is just showing its face, we recognize it right away and deal with it before it becomes too strong. If you know someone to be a pickpocket, then you will soon spot that person even if he mingles with a crowd of 20 or 30 people, and you will keep a careful eye on him so he will not be able to steal your bag.

Wolf: The goal is then to enhance your sensitivity to the subtle flow of your emotions to be able to control them before they become a menace.

Matthieu: Yes, by becoming more and more familiar with the workings of the mind and cultivating mindfulness of the present moment, you will not let the spark of afflictive emotions become like a powerful fire that blazes out of control and destroys your happiness and that of others. In the beginning, this requires purposeful effort. Later, it can become effortless.

Wolf: It is not unlike a scientific endeavor except that the analytical effort is directed toward the inner rather than the outer world. Science also attempts to understand reality by increasing the resolving power of instruments, training the mind to grasp complex relations, and decomposing systems into ever-smaller components.

Matthieu: It is said in the Buddhist teachings that there is no task so difficult that it cannot be broken down into a series of small, easy tasks.

Wolf: Your object of inquiry appears to be the mental apparatus and your analytical tool, introspection. This is an interesting self-referential approach that differs from the Western science of mind because it emphasizes the first-person perspective and collapses, in a sense, the instrument of investigation with its object. The Western approach, while using the first-person perspective for the definition of mental phenomena, clearly favors the third-person perspective for its investigation. I am curious to find out whether the results of analytical introspection match those obtained by cognitive neuroscience. Both approaches obviously try to develop a differentiated and realistic view of cognitive processes. It may be that our Western way of using introspection is not sophisticated enough. The fact is that some concepts of the human brain's organization that have been derived from intuition and introspection are in striking conflict with concepts derived from scientific inquiry— which sometimes gives rise to heated debates between neuroscientists and scholars of the humanities. What guarantees that the introspective technique for the dissection of mental phenomena is reliable? If it is the consensus among those who consider themselves experts, how can you compare and validate subjective mental states? There is nothing another person can look at and judge as valid; the observers can only rely on the verbal testimony of subjective states.

GRADUAL AND LASTING CHANGES

Matthieu: It is the same with scientific knowledge. You first have to rely on the credible testimony of a number of scientists, but later you can train in the subject and verify the findings firsthand. This is quite similar to contemplative science. You first need to refine the telescope of your mind and the methods of investigations for years to find out for yourself what other contemplatives have found and all agreed on. The state of pure consciousness without content, which might seem puzzling at first sight, is something that all contemplatives have experienced. So it is not just some sort of Buddhist dogmatic theory. Anyone who takes the trouble to stabilize and clarify his or her mind will be able to experience it.

Regarding cross-checking interpersonal experience, both contemplatives and the texts dealing with the various experiences a meditator might encounter are quite precise in their descriptions. When a student reports on his inner states of mind to an experienced meditation master, the descriptions are not just vague and poetic. The master will ask precise questions and the student replies, and it is quite clear that they are speaking about something that is well defined and mutually understood.

However, in the end, what really matters is the way the person gradually changes. If, over months and years, someone becomes less impatient, less prone to anger, and less torn apart by hopes and fears, then the method he or she has been using is a valid one. If it becomes inconceivable for someone to willingly harm another person, if the person has gradually developed the inner resources to successfully deal with the ups and downs of life, then real progress has occurred. It is said in the teachings that it is easy to be a great meditator when sitting in the sun with a full belly, but meditators are truly put to the test when faced with adverse circumstances. That is the time when you will really measure the change that has occurred in your way of being. When you are confronted with someone who criticizes or insults you, if you don't blow a fuse but know how to deal skillfully with the person while maintaining your inner peace, you will have achieved some genuine emotional balance and inner freedom. You will have become less vulnerable to outer circumstances and your own deluded thoughts.

An ongoing study seems to indicate that while they are engaged in meditation, practitioners can clearly distinguish, like everyone who is not distracted, between pleasant and aversive stimuli, but they react much less emotionally than control subjects. While retaining the capacity of being fully aware of something, they succeed in not being carried away by their emotional responses.[1] Normal subjects either do not perceive the stimuli (e.g., when being purposely distracted by being asked to perform a cognitively demanding task) and do not react or perceive it and react strongly.

Wolf: I can see the virtue of this attitude. However, negative emotions also have important functions for survival. They have not evolved and been conserved by chance; they help us to survive. They protect us and help us avoid adverse situations. We have only talked about the disconnection and detachment of the negative components while preserving the positive components—empathy, love, carefulness, mindfulness, and diligence. For reasons of symmetry, one should expect that positive emotions also hamper an unbiased view of the world and fade with mental training.

Matthieu: If love and empathy are biased with attachment and grasping, then they will surely be accompanied by a distortion of reality. Consequently, from a Buddhist perspective, biased empathy and grasping love are not positive because they result in suffering. Conversely, altruistic love has positive effects on all concerned: the beneficiaries as well as the one who expresses that love. Similarly, strong indignation in the face of injustice can motivate one to engage energetically in actions intended to right the wrong. If such indignation is not mixed with hatred and is not superimposed on reality, then it is constructive, unlike malevolent, out-of-control anger. It will result in less suffering and greater well-being for all. The positive or negative nature of an emotion should be assessed according to its motivation—altruistic or selfish—and its consequences in terms of well-being or suffering.

Wolf: How can we conceive of a process that is uniquely initiated by our own brain? You want to change something in your brain by reducing as many intrusions as possible from outside; you can undertake a long promenade through your own brain trying to evoke certain feelings. This would seem to require a certain dissociation, a level splitting, because there needs to be an agent that works on another level to induce a change. You need to monitor your emotions, you need to alert your inner senses to have those emotions—because I think you can only work on them if you activate them—and then you have to learn to differentiate them. How do you do this? What are the tools?

OUTER AND INNER ENRICHMENT

Matthieu: The mind obviously has the ability to know and train itself. People do that all the time without calling it meditation. They voluntarily memorize things, as a student will do; they enhance their mental skills in playing chess and solving various problems through mental training. Meditation is simply a more systematic way of doing this with wisdom— that is, with an understanding of the mechanisms of happiness and suffering. This process requires perseverance. You need to train again and again. You can't learn to play tennis by holding a racket for a few minutes every few months. With meditation, the effort is aimed at developing not a physical skill but an inner enrichment. I understand that the development of brain functions comes from exposure to the outer world. If you are born blind, then the visual areas of the brain will not develop and will even be colonized by the auditory functions, which are more useful to a blind person.[2] In the late 1990s, research showed that rats kept in a plain cardboard box show reduced neuronal connectivity. But if they are placed in an amusement park for rats, with wheels, tunnels, and friends, within a month they form many new functional connections.[3] Soon after, neuroplasticity was also shown to exist throughout the life course in humans.[4] However, most of the time, our engagement with the world is semi-passive. We are exposed to something and react to it, thus increasing our experience. We could describe this process as an outer enrichment.

In the case of meditation and mind training, the outer environment might change only minimally. In extreme cases, you could be in a simple hermitage in which nothing changes or sitting alone always facing the same scene day after day. So the outer enrichment is almost nil, but the inner enrichment is maximal. You are training your mind all day long with little outer stimulation. Furthermore, such enrichment is not passive, but voluntary, and methodically directed.

When you engage for eight or more hours a day in cultivating certain mental states that you have decided to cultivate and that you have learned to cultivate, you reprogram the brain.

Wolf: In a sense, you make your brain the object of a sophisticated cognitive process that is turned inward rather than outward toward the world around you. You apply the cognitive abilities of the brain to studying its own organization and functioning, and you do so in an intentional and focused way, similar to when you attend to events in the outer world and when you organize sensory signals into coherent percepts. You assign value to certain states and you try to increase their prevalence, which probably goes along with a change in synaptic connectivity in much the same way as it occurs with learning processes resulting from interactions with the outer world.[5]

Let us perhaps briefly recapitulate how the human brain adapts to the environment because this developmental process can also be seen as a modification or reprogramming of brain functions. Brain development is characterized by a massive proliferation of connections and is paralleled by a shaping process through which the connections being formed are either stabilized or deleted according to functional criteria, using experience and interaction with the environment as the validation criterion.[6] This developmental reorganization continues until the age of about 20. The early stages serve the adjustment of sensory and motor functions, and the later phases primarily involve brain systems responsible for social abilities. Once these developmental processes come to an end, the connectivity of the brain becomes fixed, and large-scale modifications are no longer possible.

Matthieu: To some extent.

Wolf: To some extent, yes. The existing synaptic connections remain modifiable, but you can't grow new long-range connections. In a few distinct regions of the brain, such as the hippocampus and olfactory bulb, new neurons are generated throughout life and inserted into the existing circuits, but this process is not large scale, at least not in the neocortex, where higher cognitive functions are supposed to be realized.[7]

Matthieu: A study of people who have practiced meditation for a long time demonstrates that structural connectivity among the different areas of the brain is higher in meditators than in a control group.[8] Hence, there must be another kind of change allowed by the brain.

PROCESSES OF NEURONAL CHANGES

Wolf: I have no difficulty in accepting that a learning process can change behavioral dispositions, even in adults. There is ample evidence of this from reeducation programs, where practice leads to small but incremental behavior modifications. There is also evidence for quite dramatic and sudden changes in cognition, emotional states, and coping strategies. In this case, the same mechanisms that support learning—distributed changes in the efficiency of synaptic connections—lead to drastic alterations of global brain states. The reason is that in a highly nonlinear, complex system such as the brain, relatively small changes in the coupling of neurons can lead to phase transitions that can entrain radical alterations of system properties. This can occur in association with traumatic or cathartic experiences. The rare sudden onset of psychosis is also likely due to such global state changes, but this is probably not what occurs with meditation because this practice seems to lead to slow changes.[9]

Matthieu: You could also change the flow of neuron activity, as when the traffic on a road increases significantly.

Wolf: Yes. What changes with learning and training in the adult is the flow of activity. The fixed hardware of anatomical connections is rather stable after age 20, but it is still possible to route activity flexibly from A to B or from A to C by adding certain signatures to the activity that ensure that a given activation pattern is not broadcast in a diffuse way to all connected brain regions but sent only to selected target areas. The strength of interactions among centers can be modified by actually modulating the efficiency of the connecting synapses or dynamically configuring virtual highways. The latter strategy is probably based on the same principle as the tuning of a receiver to a specific radio station. The receiver is entrained into the same oscillation frequency as the sender.[10] In the brain, myriad senders are active all the time. Their messages must be selectively directed to specific targets, and this routing must occur in a task-dependent way. Thus, different functional networks need to be configured from moment to moment, and this must be achievable at time scales much faster than the learning-dependent changes of synaptic efficacy. The training phase in meditation is probably capitalizing on the slow, learning-related modifications of synaptic efficiency, whereas the fast engagement in a particular meditative state of which experts seem to be capable likely relies on more dynamic routing strategies.

Matthieu: You could thus gradually slow down the traffic on pathways of hatred and open wide the routes of compassion, for instance. So far, the results of the studies conducted with trained meditators indicate that they have the faculty to generate clean, powerful, well-defined states of mind, and this faculty is associated with some specific brain patterns. Mental training enables one to generate those states at will and to modulate their intensity, even when confronted with disturbing circumstances, such as strong positive or negative emotional stimuli. Thus, one acquires the faculty to maintain an overall emotional balance that favors inner strength and peace.

Wolf: So you have to use your cognitive abilities to identify more clearly and delineate more sharply the various emotional states, and to train your control systems, probably located in the frontal lobe, to increase or decrease selectively the activity of subsystems responsible for the generation of the various emotions.

Matthieu: You can surely refine your knowledge of the various aspects of mental processes themselves.

Wolf: Sure. You are aware of them, and you can familiarize yourself with them by focusing attention on them and then differentiating between them, forming category boundaries as one does when perceiving the outer world.

Matthieu: You can also identify the mental processes that lead to suffering and distinguish them from those that contribute to well-being, those that feed mental confusion, and those that preserve lucid awareness.

Wolf: Another analogy for this process of refinement could be the improved differentiation of objects of perception, which is known to depend on learning. With just a little experience, you are able to recognize an animal as a dog. With more experience, you can sharpen your eye and become able to distinguish with greater and greater precision dogs that look similar. Likewise, mental training might allow you to sharpen your inner eye for the distinction of emotional states. In the naïve state, you are able to distinguish good and bad feelings only in a global way. With practice, these distinctions would become increasingly refined until you could distinguish more and more nuances. The taxonomy of mental states should thus become more differentiated. If this is the case, then cultures exploiting mental training as a source of knowledge should have a richer vocabulary for mental states than cultures that are more interested in investigating phenomena of the outer world.

EMOTIONAL NUANCES

Matthieu: Buddhist taxonomy describes 58 main mental events and various subdivisions thereof. It is quite true that by conducting an in-depth investigation of mental events, one becomes able to distinguish increasingly more subtle nuances. If you look at a painted wall from a distance, it looks quite homogenous. However, if you look closely, you will see many imperfections: the surface is not as smooth as it seems; it has bumps and holes and white, yellowish, and dark spots, and so on. Similarly, when we look closely at our emotions, we find that they have many different aspects. Take anger, for instance. Often anger can have a malevolent component, but it can also be rightful indignation in the face of injustice. Anger can be a reaction that allows us to rapidly overcome an obstacle preventing us from achieving something worthwhile or remove an obstacle threatening us. However, it could also reflect a tendency to be short-tempered.

If you look carefully at anger, you will see that it contains aspects of clarity, focus, and effectiveness that are not harmful in and of themselves. Likewise, desire has an element of bliss that is distinct from attachment; pride has an element of self-confidence that does not lapse into arrogance; and envy entails a drive to act that, in itself, is not yet deluded, as it will later become when the afflictive state of mind of jealousy sets in.

So if you are able to recognize those aspects that are not yet negative and let your mind remain in them, without drifting into the destructive aspects, then you will not be troubled and confused by these emotions. This process is not easy, to be certain, but one can cultivate this capacity through experience.

EFFORTLESS SKILLS

Matthieu: Another result of cultivating mental skills is that, after a while, you will no longer need to apply contrived efforts. You can deal with the arising of mental perturbations like the eagles I see from the window of my hermitage in the Himalayas. The crows often attack them, even though they are much smaller. They dive at the eagles from above trying to hit them with their beaks. However, instead of getting alarmed and moving around to avoid the crow, the eagle simply retracts one wing at the last moment, letting the diving crow pass by, and extends its wing back out. The whole thing requires minimal effort and is perfectly efficient. Being experienced in dealing with the sudden arising of emotions in the mind works in a similar way. When you are able to preserve a clear state of awareness, you see thoughts arise; you let them pass through your mind, without trying to block or encourage them; and they vanish without creating many waves.

Wolf: That reminds me of what we do when we encounter severe difficulties that require fast solutions, such as a complicated traffic situation. We immediately call on a large repertoire of escape strategies that we have learned and practiced, and then we choose among them without much reasoning, relying mainly on subconscious heuristics. Apparently, if we are not experienced with contemplative practice, we haven't gone through the driving school for the management of emotional conflicts. Would you say this is a valid analogy?

Matthieu: Yes, complex situations become greatly simplified through training and the cultivation of effortless awareness. When you learn to ride a horse, as a beginner you are constantly preoccupied, trying not to fall at every movement the horse makes. Especially when the horse starts galloping, it puts you on high alert. But when you become an expert rider, everything becomes easier. Riders in eastern Tibet, for instance, can do all kinds of acrobatics, such as shooting arrows at a target or catching something on the ground while galloping at full speed, and they do all that with ease and a big smile on their face.

One study with meditators showed that they can maintain their attention at an optimal level for extended periods of time. When performing what is called a *continuous performance task*, even after 45 minutes, they did not become tense and were not distracted even for a moment.[11] When I did this task myself, I noticed that the first few minutes were challenging and required some effort, but once I entered a state of "attentional flow," it became easier.

Wolf: This resembles a general strategy that the brain applies when acquiring new skills. In the naïve state, one uses conscious control to perform a task. The task is broken down into a series of subtasks that are sequentially executed. This requires attention, takes time, and is effortful. Later, after practice, the performance becomes automatized. Usually, the execution of the skilled behavior is then accomplished by different brain structures than those involved in the initial learning and execution of the task. Once this shift has occurred, performance becomes automatic, fast, and effortless and no longer requires cognitive control. This type of learning is called *procedural learning* and requires practice. Such automatized skills often save you in difficult situations because you can access them quickly. They can also often cope with more variables simultaneously due to parallel processing. Conscious processing is more serialized and therefore takes more time. Do you think you can apply the same learning strategy to your emotions by learning to pay attention to them, differentiate them, and thereby familiarize yourself with their dynamics so as to later become able to rely on automatized routines for their management in case of conflict?

Matthieu: You seem to be describing the meditation process. In the teachings, it says that when one begins to meditate, on compassion, for instance, one experiences a contrived, artificial form of compassion. However, by generating compassion over and over again, it becomes second nature and spontaneously arises, even in the midst of a complex and challenging situation. Once compassion becomes truly part of your mind stream, you don't have to make special efforts to sustain it. We say it's "meditating without meditation": you are not actively "meditating," but at the same time you are never separated from meditation. You simply dwell effortlessly and without distraction in this wholesome, compassionate state of mind.

Wolf: It would be really interesting to look with neurobiological tools at whether you have the same shift of function that you observe in other cases where familiarization through learning and training leads to the automation of processes. In brain scans, one observes that different brain structures take over when skills that are initially acquired under the control of consciousness become automatic.

Matthieu: That is what a study conducted by Julie Brefczynski and Antoine Lutz at Richard Davidson's lab seems to indicate. Brefczynski and Lutz studied the brain activity of novice, relatively experienced, and very experienced meditators when they engage in focused attention. Different patterns of activity were observed depending on the practitioners' level of experience. Relatively experienced meditators (with an average of 19,000 hours of practice) showed more activity in attention-related brain regions compared with novices. Paradoxically, the most experienced meditators (with an average of 44,000 hours of practice) demonstrated less activation than the ones without as much experience. These highly advanced meditators appear to acquire a level of skill that enables them to achieve a focused state of mind with less effort. These effects resemble the skill of expert musicians and athletes capable of immersing themselves in the "flow" of their performances with a minimal sense of effortful control.[12] This observation accords with other studies demonstrating that when someone has mastered a task, the cerebral structures put into play during the execution of this task are generally less active than they were when the brain was still in the learning phase.

Wolf: This suggests that the neuronal codes become sparser, perhaps involving fewer but more specialized neurons, once skills become highly familiar and are executed with great expertise. To become a real expert seems to require then at least as much training as is required to become a world-class violin or piano player. With four hours of practice a day, it would take you 30 years of daily meditation to attain 44,000 hours. Remarkable!

RELATING TO THE WORLD

Matthieu: Mind training leads to a refined understanding of whether a thought or an emotion is afflictive, attuned to reality or based on a completely distorted perception of reality.

Wolf: What is the difference between the two? You consider the afflictive state as enslaving, as narrowing, as masking valid cognition—in brief, as a fundamentally negative state that is not tuned to reality. I fully understand that your strategy works well as long as the source of conflict is solely your own pathology, but most conflicts arise from interactions with the world, which is clearly not free of conflict. Are you not assuming that the world is ideal and good and that it would be sufficient to purify one's mind to be able to recognize this fact?

Matthieu: There are two ways of looking at this. The first one is to clearly recognize the flaws and shortcomings of the world, where beings are mostly ruled by mental confusion, obscuring emotions, and suffering. The other way is to recognize that each and every sentient being in this world has the potential to get rid of such afflictions and actualize wisdom, compassion, and other such qualities.

Afflictive mental states begin with self-centeredness, with increasing the gap between self and others, between oneself and the world. They are associated with an exaggerated feeling of self-importance, an inflated self-cherishing, a lack of genuine concern for others, unreasonable hopes and fears, and compulsive grasping toward desirable objects and people. Such states come with a high level of reality distortion. One solidifies outer reality and believes that the good or bad, desirable or undesirable qualities of outer things intrinsically belong to them instead of understanding that they are mostly projections of our mind.

In contrast, an act of unconditional benevolence, of pure generosity— as when you do something to make a child happy, help someone in need, save a life even, with no strings attached—even if nobody knows what you have done, this generates deep satisfaction and fulfillment.

Wolf: I am fascinated by the fact that what you tell me seems to put strong emphasis on the cultivation of an autonomous self. Not a selfish, possessive ego, but a strong, confident self.

Matthieu: I am not talking about the strength of the ego or self-centeredness, which is the troublemaker, but a deep sense of confidence that comes from having gained some knowledge about the inner mechanisms of happiness and suffering, from knowing how to deal with emotions, and thus from having gathered the inner resources to deal with whatever comes your way.[13]

HOW YOUNG CAN ONE START TO MEDITATE?

Wolf: I take from your description that meditation requires a high level of cognitive control. However, cognitive control depends on the prefrontal cortex, which becomes fully functional only during late adolescence. Does this imply that only adults can practice meditation? If not, would it not be preferable to begin with meditation as early as possible to capitalize on the plasticity of the brain and make it an integral part of education? We know that the acquisition of other abilities, such as playing the violin or learning a second language, is much easier in early life. Can children master a technique that requires so much cognitive control?

Matthieu: Indeed there are stages in our emotional development, but I think that even at early stages, there is a possibility to do some kind of training. In our monastery at Shechen, we don't formally teach meditation to children and young novices (from 8 to 14 years old). But they do participate in long ceremonies in the temple, which resemble group meditations, during which there is a soothing atmosphere of inner calm and emotional rest, so the children begin to be exposed to these states of mind at an early age. I am sure it helps a lot to simply provide an environment that calms the mind rather than constantly provoking waves of emotional disturbances, as is often the case in the West, with noise, violence on TV, video games, and the like.

Besides this, in a traditional Buddhist setting, young children are mostly taught through example. They see their parents and educators behave on the basis of the principles of nonviolence toward humans, animals, and the environment. One cannot underestimate the strength of emotional contagion, as well as the way of being's contagion. One's inner qualities are immensely influential on those who share one's life. One of the most important things is to help children become skilled in identifying their emotions and those of others, and to show them basic ways of dealing with emotional outbursts.

Wolf: This is one of the goals of every educational system, to strengthen the ability to control one's emotions, and a rich repertoire of tools is available to achieve this: reward and punishment, creating attachment to role models, educational games, storytelling, and so on. All cultures have recognized the virtues of controlling emotions and developed a large variety of educational strategies to that end.

Matthieu: I must add that, although it certainly requires some maturity to achieve lasting stability in emotional control, it still seems possible to begin this process at an early age. Children do find strategies to recover a sense of balance and inner peace after going through emotional upheaval. In a book called *The Joy of Living*, Mingyur Rinpoche recounts how as a child he was extremely anxious and had frequent panic attacks. He was then living in Nubri, in the mountains of Nepal, near the Tibetan border. He came from a nice, loving family—his grandfather and father were great meditators—and did not experience any particular traumatic event, but he had these uncontrollable bursts of inner fear. But even at the age of six or seven years old, he found a way to alleviate his panic attacks. He used to go to a cave nearby and sit there alone, meditating in his own way for a couple of hours. He felt a welcome sense of peace and relief, as if turning off the heat, and he deeply appreciated the quality of those contemplative moments. Still, that was not enough to get rid of his anxiety, which kept on creeping back.

At the age of 13, he felt a strong aspiration to do a contemplative retreat and embarked on the traditional three-year retreat that is often practiced in Tibetan Buddhism. In the beginning, things became even worse. So one day he decided that enough was enough and that the time had come to use all the teachings he had received from his father to go to the depth of his problem. He meditated for three days uninterruptedly, not coming out of his room, looking deep into the nature of mind. At the end of it, he had gotten rid of his anxiety forever. When you now meet this incredibly kind, warm, and open person, who radiates well-being and inner peace, displays such great warmth and sense of humor, and teaches with limpid clarity on the nature of mind, you find it hard to believe that he ever experienced anything close to anxiety. He is a living testimony of the power of mind training and furthermore of the possibility to embark on it from an early age.[14]

MENTAL DISTORTIONS

Wolf: In German we have a saying, "*Komm zu dir*," which means "cut the strings"—the ties that attach you to something, that make you do what others want, that make you believe what others believe, that make you be kind because somebody else wants you to be kind. If you get caught in this net of dependencies, then we say that you "lose yourself." This is why a protective environment that generously grants self-determination is indispensable, as long as the cognitive control mechanisms of children are strong enough to protect them from losing themselves in the face of imposed intrusions and expectations.

Matthieu: After recovering from a fit of anger, we often say, "I was beside myself" or "I wasn't myself."

Wolf: "*Ich war außer mir*"—"I was out of myself"—we also say this in German. Life sometimes confronts us with situations that we simply cannot cope with by remaining equanimous and that drive us "out of ourselves." However, we have developed strategies to recover equilibrium. Some of these may be innate, whereas others may be acquired by learning.

Matthieu: This is meant to be the fruit of practice. Strong emotions may still arise in the mind, but instead of invading and overpowering the mind, they vanish like a whisper.

Wolf: That sounds wonderful. Usually it takes quite some time until you get back into a quiet state. One of the reasons is that the stress hormones released in highly aversive situations decay rather slowly.

Matthieu: With experience it does not necessarily take a long time. In fact, it can come down as quickly as bubbling milk taken off the fire. If you let the emotion, even a strong one, pass through your mind without fueling it, without letting the spiral of thoughts spin out of control, the emotion will not last and will vanish by itself.

ATTENTION AND COGNITIVE CONTROL

Wolf: Earlier, we were talking about the possibility of using mental practice as a tool to fine-tune the inner eye and the ability to use introspection to explore the cognitive functions of the brain and learn to form more differentiated categories of emotions and of cognitive processes in the same way that one can fine-tune one's perception of the outer world. Experts in perfume factories, the "noses," learn through practice to distinguish mixtures of odors that for most of us smell the same. It is conceivable that mental practice can do the same thing to the cognitive abilities of the brain and sharpen awareness of one's own cognitive processes. This does require a substantial amount of cognitive control because in this case attention—unlike in the case of the "noses"—has to be directed toward processes originating within the brain.

There is now convincing neurobiological evidence suggesting that mental practice uses attention mechanisms to activate and analyze internal processes so that they can become the subject of learning processes.[15] I allude to the seminal work by Richard Davidson and Antoine Lutz, who recorded electroencephalograms of you and other Buddhist practitioners while you were meditating.[16] When I first saw these data at the meeting in Paris that was organized in memoriam of Francisco Varela, a good friend of both of ours, I was struck by the fact that there was a striking increase in meditators' brains of the amplitude of oscillatory activity in a frequency range of 40 Hz, the so-called gamma frequency band. These oscillations were discovered some 25 years ago in the visual cortex and were suspected to play an important role in cognitive processes. Since then much work has been performed to investigate the putative functions of oscillations and synchrony in neuronal processing.

Of the many different functions that this temporal patterning of neuronal activity is likely to serve, its involvement in attentional mechanisms is particularly important in the present context. Several laboratories provided evidence that focused attention is associated with an enhancement of gamma oscillations and neuronal synchrony.[17] If attention is directed to a particular subsystem in the brain to prepare it for processing, one observes an increase of synchronous gamma oscillations in that system. If you are about to direct your attention to a visual object, then the anticipation of having to process signals from this visual object increases oscillatory activity in the beta and gamma frequency range in visual areas of the cerebral cortex.

Likewise, if one anticipates that one will have to process an auditory signal, and one will have to use this signal to initiate a motor act, the brain begins to synchronize the oscillatory activity among the areas that will be involved in the future process—in this case, the auditory cortex and the premotor and motor areas. This facilitates rapid "handshaking" between the concerned areas and prepares the necessary coordination between sensory and executive structures.[18]

Thus, when a stimulus actually appears, the responses to this stimulus are enhanced and better synchronized than when the stimulus was not anticipated. This prerequisite is necessary to ensure rapid information processing and safe transmission of computational results across the cortical network.[19]

The phenomenon of binocular rivalry illustrates the close relations among synchronous oscillatory activity, conscious perception, and attention. If the two eyes are presented with different patterns that cannot be fused into a single coherent percept, then only one of the two images is perceived at any one time. If, for instance, a set of vertical lines is shown to the right eye and a set of horizontal lines to the left eye, one does not perceive a superposition of the two gratings, which would look like a checkerboard. Rather one sees either the vertical or horizontal grating, and these percepts keep on alternating every few seconds due to internal switching mechanisms. The question is, how is this selection and switching process achieved at the neuronal level?

At the early stages of visual processing in the primary visual cortex, this switch in perception is associated with a change in the synchronization of neuronal responses to the gratings. The grating that is actually perceived at a particular moment evokes responses that are more synchronized in the 40-Hz range than the responses to the grating not perceived at that moment.[20] Each eye physically "sees" the same pattern all the time, but the subject perceives only the vertical or horizontal grating. These experiments suggest that it is easier for perceptual signals to access the level of conscious processing if they are well synchronized.

Matthieu: Why does this switching happen without the subject being able to control it?

Wolf: The signals from either the right or left eye are suppressed to avoid seeing double images. We perform this suppression all the time without being aware of it, and it is only under experimental conditions that we take notice of this phenomenon of interocular suppression. Because it involves an internal process to decide which of the available sensory signals should have access to consciousness, interocular suppression is frequently used as a paradigm to investigate the signatures of neuronal activity that are required for any neuronal activity to reach the level of conscious perception. In this context, it is noteworthy that practitioners of meditation can deliberately slow down the alternation rate of binocular rivalry.[21] I experienced this myself after a few days of Zen practice while staring at the white wall in front of me. As I could infer from the changes in the far periphery of the visual field, the signals conveyed by my two eyes to my brain became suppressed in alternation at a remarkably slow rhythm of a few seconds.

Matthieu: I did that once with Brent Field at Anne Treisman's laboratory at Princeton and found out that it was possible to slow down the automatic switching between the left and right images and keep the perception of only one image up to 30 seconds or even a minute.

Wolf: The neuronal correlate of a conscious perception compared with nonconscious processing appears to be a sudden and strong increase of precise phase synchrony—or, one could also say, of coherence of oscillatory activity, first in the gamma frequency range and subsequently during the maintenance period also in lower frequency ranges. Access to consciousness seems to require a particularly well-ordered global state of the brain.[22]

Matthieu: So it's the same as Francisco Varela's "moony face experiment"?

Wolf: Yes, it is closely related to Francisco's experiment with the moony faces. He found an increase of gamma oscillations and synchrony between cortical areas, when subjects were able to identify a human face in pictograms consisting of black-and-white contours. If they failed to see a face and just perceived noninterpretable contours, the gamma oscillations had smaller amplitude and were less well synchronized.[23]

Now, this excursion was long but necessary, as you will see that it is closely related to the neuronal correlates of meditation—what Richard Davidson saw in your brain when you engaged in meditation.

Matthieu: Not only me of course, but quite a few other meditators...

Wolf: ...quite a few others, fortunately, because in science you need repeatability. What he saw was a surprising increase of a highly coherent oscillatory activity in the gamma frequency range of 40 to 60 Hz.

The most interesting observation, however, was that this increase occurred over the central and frontal brain regions but not, as is the case when you direct attention to the outer world, over sensory areas. This finding suggests that you engaged your attentional mechanism to focus attention on processes in higher cortical areas, those areas that process highly abstract concepts, symbols, and maybe also feelings and emotions. It is difficult with electroencephalographic investigation to localize the activated areas, but the source of this activity is likely in areas other than the primary sensory areas because there was no sensory stimulation. Dangerously misleading in such measurements are artifacts caused by non-neuronal processes, such as muscle contractions and eye movements. I hope these potential sources of artifacts have been controlled for in the experiments on meditators.

One way to interpret these findings is that you intentionally activate internal representations, focus your attention on them, and then work on them in much the same way as you process external information. You apply your cognitive abilities to internal events.

Matthieu: Or you keep a meta-awareness of a particular state that you are trying to develop, such as compassion, and maintain this meditation state moment after moment—

Wolf: —keeping your attention focused on particular internal states, which can be emotions or the contents of imagination. In essence, it is the same strategy as one applies with the perception of the outer world—except that most of us are far less familiar with focusing attention on inner states.

Matthieu: This fits with the definition of meditation, which is to cultivate a particular state of mind without distraction. Two Asian words are usually translated as "meditation": in Sanskrit, *bhavana* means "to cultivate," and in Tibetan, *gom* means "to become familiar with something that has new qualities and insights as well as a new way of being." So meditation cannot be reduced to the usual clichés of "emptying the mind" and "relaxing."

Wolf: Just as in cases where we focus attention on external events, learning occurs as a consequence of attending to something. When one attentively observes an object, one learns about the object. Changes in synaptic connections between neurons occur, and the next this object is observed, it will appear more familiar. It is recognized much more easily and faster, and it can be recalled from memory and enter awareness—but all this is only possible if one directs one's attention toward the object while it is perceived.

Matthieu: It could be maintaining and cultivating the experience of benevolence. Altruistic love, for instance, occurs in everyone's mind from time to time, but it usually does so in a transient way and is quickly replaced by another state of mind. Because we do not cultivate altruistic love systematically, this short-lived state will not be well integrated in the mind and will not lead to lasting changes in our dispositions. We all experience thoughts of lovingkindness, generosity, inner peace, and freedom from conflict. Yet these thoughts are fleeting and will soon be replaced by other thoughts, including afflictive ones such as anger and jealousy. To fully integrate altruism and compassion in our mind stream, we need to cultivate them over longer periods of time. We need to bring them to our minds and then nurture them, repeat them, preserve them, and enhance them, so that they gradually fill our mental landscape in a more durable way.

The idea is not only to generate but to perpetuate over an extended period of time a powerful state of mind that is saturated with benevolence. Elements of repetition and perseverance are common to all forms of training. However, the particularity here is that the skills you are developing are fundamental human qualities such as compassion, attention, and emotional balance.

Wolf: Right. Meditation, then, is a highly active, attentive process. By focusing attention on those internal states, you familiarize yourself with them, you get to know them, and this facilitates recall if you want to activate them again.

This must go along with lasting changes at the neuronal level. Any activity in the brain that is occurring under the control of attention is memorized. There are modifications in synaptic transmission; synapses will strengthen or weaken. This in turn will lead to changes in the dynamical state of neuronal assemblies. Thus, through mental training, you create novel states of your mind, and you learn to retrieve them at will. I find it remarkable that this possibility has been discovered at all. What was the incentive to withdraw attention from the outer world, direct it toward internal states, subject them to cognitive dissection, and eventually gain control over them? Why is it that Eastern traditions have focused so much on the internal rather than the external universe?

Matthieu: Well, I guess it is because these mental states are key determinants of happiness and suffering. This is truly important in anyone's life. What I find even more surprising is how little attention has been paid in the Western world to the inner conditions of well-being and how much people underestimate the capacity of the mind to transform the way we experience things.

Wolf: A particularly fascinating aspect is that this kind of mental training leads to changes in the brain that are long lasting and persist beyond the meditation process. A recent study from Harvard University showed that in long-term meditators, the volume of the cerebral cortex is increased in certain areas of the brain.[24] Research done at my daughter Tania's laboratory has also shown that structural changes occur in the brain of subjects initially new to meditation who trained for nine months in three types of practices: three months of mindfulness, three months in perspective taking, and three months in lovingkindness. Each type of meditation produced structural modification in a specific area, which are different from one meditation to the other. Such increases in volume have also been observed after learning motor skills or intensive sensory stimulation and are due to a learning-dependent increase of the neuropile (i.e., the compartment containing the connections between neurons). The number and size of synapses and their targets, the spines of the dendrites, increase, just as with other forms of training and learning.[25]

ATTENTIONAL BLINK

Wolf: Another well-controlled study points in the same direction and suggests long-term modifications of the mechanisms that control attention. It appears that the maintenance of the high level of attention required to sustain meditative states causes a modification of the mechanisms that sustain attention.

Let me explain the finding. A researcher in the laboratory of Anne Treisman, an expert in attention research, investigated a phenomenon called *attentional blink* in long-term meditators.[26] One can show a sequence of words or images in rapid succession and adjust the parameters in such a way that subjects perceive only a fraction of the stimuli. Once subjects perceive one image, the next image, maybe even the next two images, will not be perceived because the brain is still engaged with the processing of the first image and thus has no attentional resources left to process the following image or images. This inability to process the subsequent images is called attentional blink. The idea is that attention, while it is engaged in processing one consciously perceived image, is not available for the processing of the next one. Heleen Slagter and Antoine Lutz have also shown that after three months of intensive training in meditation on full awareness, attentional blink was considerably reduced.[27]

Matthieu: So, when you have a quick succession of images, letters, or words, when you clearly identify one of them, that process involves your mind to such an extent that you will not be able to see one or more of the images that follow just after the one you have recognized.

Wolf: The time interval during which you are "blind" is in the range of 50 to a few hundred milliseconds depending on the complexity of the processed image and subject age. The surprising finding was that experienced meditators, even if they had already reached a certain age—the blink interval increases with age because attentional mechanisms slow down—had remarkably short blink intervals. They perceived each of the stimuli even at high presentation rates.

Matthieu: There is an unpublished result about a 65-year-old meditator who showed no attentional blink at all.

Wolf: We have confirmed this. In aged long-term practitioners, the attentional blink was as short as in young controls.[28] This finding indicates that long-term meditation alters attentional mechanisms. Another remarkable finding has been reported in the follow-up study that Richard Davidson performed with you and your colleagues. It showed a close correlation between the amplitude of the attention-related, highly synchronous gamma oscillations over central cortical areas and your subjective judgment on the actual depth and clarity of your meditative state. It is important to demonstrate such correlations between biophysical measures and subjective phenomena; if there is a statistically significant correlation, then it is likely that there is more than an accidental coincidence and perhaps even a causal relation. As far as I know, these robust and convincing data show that meditation is associated with a special brain state and does have lasting effects on brain functions.

Matthieu: Regarding attentional blink, from an introspective perspective, it would seem that usually someone's attention is captured by the object because it goes to the object, sticks to the object, and then has to disengage from the object. There is a moment of thinking, "Oh, I have seen a tiger" or "I have seen that word." Then it takes some time to let it go. But, if you simply remain in the state of open presence, which is the state that works best to reduce attentional blink, you simply witness the image without attaching to it and therefore without having to disengage from it. When the next image flashes, a 20th of a second later, you are still there, ready to perceive it.

Wolf: So the process of meditation has two effects: You learn to work on your own attentional mechanisms, and then you become an expert in engaging and disengaging attention at will. The question is how deeply these practitioners process the individual pictures. Apparently they attach less attention to each image and therefore can perceive the successive images more easily. Could it be that they just process less thoroughly and therefore can follow more rapidly than naïve subjects, that they perform less analysis and therefore are less refractory? Are meditators in general dealing with the phenomena in the outer world in a different, perhaps more superficial way, just brushing past it and not taking anything seriously?

Matthieu: I don't think it's a question of being "serious" but rather of the relative magnitude of grasping and attaching to perceptions and outer phenomena.

Wolf: Not attaching to them?

Matthieu: Yes. Buddhism says that if we don't engage constantly in the process of attraction and repulsion, this is liberating. You also spoke about instruments, such as microscopes and telescopes, with which humans extend their cognition. From a contemplative perspective, fine tuning one's introspection toward perceptive and mental processes, rather than being powerless against and blindly caught in their automatisms, corresponds to enhancing the quality and power of the mind's telescope. This allows one to see those processes happening in real time and not be carried away and fooled by them.

It seems that the different types of meditation that have been investigated have all had quite different signatures in the brain. They might all generate gamma waves, with different magnitudes, but they certainly activate distinct areas of the brain.

Wolf: This is what you would expect because if you direct your attention toward particular emotions, train in developing compassion, or train in pure attention and empty the workspace of consciousness of any other content, then you are probably engaging different brain systems, which should result in different activation patterns. You will probably always find the attention-dependent activation patterns because meditation always requires focused attention, but the content-related activation patterns will depend on whether you direct your attention to visual, emotional, or social contents. In addition, one expects to find specific activation patterns in the respective brain regions. The common denominator of meditation, and this may sound surprising, is the high level of cognitive control.

Matthieu: After the 2000 Mind and Life meeting, I went to visit Paul Ekman, the world's leading expert on the facial expression of emotions, at his lab in San Francisco. He had a few of us go through a test in which we were presented with faces showing a neutral expression. Then for 1/30th of a second, a picture flashed of the same face showing one of the six basic emotions that are universal to all human beings: joy, sadness, anger, surprise, fear, and disgust. You could see that there had been a change, but it was very, very fast. If you were to go slowly, frame by frame, then you would see that the single image with the emotion is clear—a broad smile, a cringe of disgust, and so on. But when it is only displayed for 1/30th of a second, it looks like only a quick twitch in the face, which then immediately goes back to its neutral expression. The emotion thus briefly displayed is quite difficult to identify without training. Those *microexpressions*, as Paul calls them, occur involuntarily all the time in daily life and are uncensored indicators of one's inner feelings, but usually we are not skillful in identifying them.

Wolf: Exactly.

Matthieu: It turns out that a small number of people naturally recognize these microexpressions quite well, and you can learn to recognize them through training. In our case, two meditators took the test. Personally, I didn't feel that I had done well on the test, and I felt that the skill required did not have much to do with meditation. It turned out, however, that we actually scored higher and were more accurate and sensitive to the microexpressions than several thousand other people tested previously.[29]

According to Paul, this capacity to identify microexpressions might have been related to an enhanced speed of cognition, which would make it easier to perceive rapid stimuli in general, or greater attunement to the emotions of other people, which would make reading them easier. In general, the ability to recognize these fleeting expressions represents an unusual capacity for accurate empathy. People who do better at recognizing these subtle emotions are more interested, curious, and open to new experiences. They are also known to be conscientious, reliable, and efficient.

Wolf: It could be related to the reduction of attentional blink and simply result from the ability to perceive short-duration events, or indeed it could indicate that your perception of emotions is more refined.

ATTENTION, RUMINATION, AND OPEN PRESENCE

Matthieu: You see, regarding attention, if your mind wanders somewhere else, when the sudden change of facial expression happens, it draws back your attention, but it's too late. The expression is already gone. But if your attention is clearly dwelling in the present moment, in a state of readiness at all times, then when the event happens, you are there. You don't have to be brought back to the present moment by a sudden change. So either it is a matter of bare attention or your sensibility or openness to others' emotion is enhanced—probably a combination of both. Now should I come back to the other points that you mentioned?

Wolf: Sure.

Matthieu: You spoke about directing attention inward. The fact is that our attention is constantly facing outward. Most of the time, we are directing our attention to the outer world, to forms, colors, sounds, tastes, smells, textures, and so forth.

Wolf: Which is important for surviving.

Matthieu: Of course, there is no question about that. If you need to cross a street, then you have to be aware of all that is going on. One of the great Tibetan masters used to face the palm of his hand outward. Then he would turn his palm inward, commenting, "Now we should look within and pay attention to what is going on in our mind and to the very nature of awareness itself." This is one of the key points of meditation. Some people find this to be a rather strange thing to do. They think it's quite unhealthy to pay too much attention to our mental processes and that we should rather remain engaged in the world. Some even find the adventure scary.

Wolf: I am going to interrupt you. Yesterday, you said this occupation with oneself, with one's inner states, is rumination and just the opposite of what meditation should do. Could you differentiate that from what you're talking about now?

Matthieu: That's different from rumination. Rumination is letting your inner chatter go on and on, letting thoughts about the past invade your mind, becoming upset again about past events, endlessly guessing the future, fueling hopes and fears, and being constantly distracted in the present. By doing so, you become increasingly disturbed, self-centered, busy, and preoccupied with your own mental fabrications and eventually depressed. You are not truly paying attention to the present moment and are simply engrossed in your thoughts, going on and on in a vicious circle, feeding your ego and self-centeredness. You are completely lost in inner distraction, in the same way that you can be constantly distracted by ever-changing outer events. This is the opposite of bare attention. Turning your attention inward means to look at pure awareness and dwell without distraction, yet effortlessly, in the freshness of the present moment, without entertaining mental fabrications.

We did some other experiments with Paul Ekman and Robert Levenson at the University of California, Berkeley, that I think relate well to this concept of bare awareness. They involved the startle reflex, which occurs, for instance, when one is confronted with a loud, surprising sound. It triggers a strong expression of surprise in the face, often a strong jerk of the body, and a significant physiological response (changes in heart rate, blood pressure, skin temperature, etc.). Like all reflexes, the startle reflects brain activity that normally lies beyond the range of voluntary regulation. Usually, the more people react, the stronger they tend to experience negative emotions, such as fear, disgust, and so on.

In our case, the scientists used a sound at the top of the threshold for auditory tolerance—a very loud explosion, like that of a gunshot or a large firecracker going off near one's ear. In general, some people do better than others at moderating the startle, but many years of studies have shown that out of the several hundred people tested, no one could prevent the muscle spasms of the face and the bodily jump. Some people almost fell off their chair and, a few seconds after the startle, displayed an expression of relief or amusement. However, when we applied the strategy of the meditation of open presence, the startle almost disappeared.[30]

Wolf: Even though you didn't know what was going to happen?

Matthieu: In some trials you can see a countdown from 10 to 1 on a screen before the explosion, and in other trials you just know that it will happen within five minutes or so. The meditator is asked to either sit in a neutral state or engage in a particular meditation state. Personally, when I used the meditation of open presence, the explosion sound seemed softer and less intrusive. *Open presence* is a state of clear awareness in which the mind is vast like the sky. The mind is not focused on anything, yet it is extremely clear and present, vivid and transparent. It is usually free from discursive thoughts, but there is no intention to block or prevent the thoughts from arising. Thoughts undo themselves as they arise, without proliferating or leaving traces. If you can remain properly in this state, the bang becomes much less disturbing. In fact, it can even enhance the clarity of the open state.

Wolf: So it's not focused attention on any content—

Matthieu: —but it's never distracted either.

Wolf: You open your window of attention—

Matthieu: —yes, but without effort. There is neither mental chatter nor particular focus of attention except resting in pure awareness, rather than focusing on it. I cannot find any better word; it is something that is luminous, clear, and stable, without grasping. That's the state of mind in which the explosion creates almost no emotional reaction in the face and no change in heart rate variability.

When we repeated the experiments on two other occasions, I tried to engage in self-induced rumination and imagination, remembering a particular vivid experience from my own life. I became completely taken up by my chain of thoughts.

Wolf: You call this *internal chatter?*

Matthieu: Yes, either internal chatter or mental fabrications. When the explosion occurred as I was purposely engaged in this distracted state,

I was much more startled by the noise. My personal interpretation is that the bang suddenly brought me back to the reality of the present moment, from which I was so far away, lost in my thoughts. However, if you remain in pure awareness, you are always in the freshness of the present moment, and the explosion is simply one of these present moments. You don't have to be brought back to anything because you are already there.

It is understandable that in normal life, when there is a surprising event that requires immediate attention, perhaps even something necessary for your survival, if you are distracted at that moment, the more your mind is wandering somewhere else, the stronger the startle will be.

Wolf: So the startle reaction would be the result of shifting attention from concrete, remembered, or presently experienced events toward the unexpected new stimulus.

Matthieu: Yes, or rather shifting from being somewhere else in your mind to the present moment.

Wolf: Thus, in a state of pure awareness, you are already in the present; attention is there but it is not directed—

Matthieu: —and yet it's available completely.

Wolf: The spotlight of attention is wide, you are prepared, you don't have to first disengage attention from some other content, and therefore you are not startled.

MINDFULNESS AND DISTRACTION

Matthieu: As we cultivate attention, we should understand that it is a powerful tool, so it should be applied to something that contributes to freedom from suffering. We can also use effortless attention to simply rest in the natural state of mind, in clear awareness that is imbued with inner peace and makes us much less vulnerable to the ups and downs of life. Whatever happens, we will not suffer much emotional disturbance and can enjoy greater stability. Obtaining this pure mindfulness of the present moment has many advantages. We may also use attention to cultivate compassion. If the mind is constantly distracted, even though it looks as if one is meditating, then the mind is powerlessly carried away all over the world like a balloon in the wind. So the increased resolution of your inner telescope, combined with sustained attention, is an indispensable tool to cultivate those human qualities that can be developed through meditation. In the end, freedom from suffering becomes a skill.

Wolf: I am fascinated by the way you put it. The fact that it requires attentional control and repetition suggests to me that you use the strategy of skill learning, which relies on procedural rather than on declarative memory. We possess two distinct mechanisms for the long-term storage of information. One is the memory system, which is able to store information acquired in a one-trial learning situation. When you bite into a piece of fruit that has a hard kernel and it hurts, you will remember this for the rest of your life and never do it again. We call this *declarative* or *episodic memory*. The contents of this memory can usually be reported verbally, we are conscious of them, and we typically store information about the event as well as the context in which it occurred and its exact timestamp in our biography.

This process is different from learning a motor skill such as piano playing, skiing, or sailing—you have to practice over and over again until you become an expert and the skill becomes automatic. This is *procedural learning* and engages *procedural memory*. You have to practice, and you have to do it in a particular way. In the beginning of skill acquisition, practice is very much under the control of attention and consciousness: you have to dissect the process into steps, and you need a teacher who tells you how to do it, or you do it by trial and error, which is less efficient—

Matthieu: —hence the importance of having a skilled teacher, especially when engaging in meditation.

Wolf: Teachers help, they speed up the process, but you have to practice yourself. The neuronal substrate that supports these skills cannot shift instantly into a new state. You have to tune the neuronal circuits little by little over a long period of time; finally, when the skill is acquired, it becomes less and less dependent on attention and becomes more and more automatized. Imagine driving your car. You don't invest any attention anymore in driving your car through a region in your city that you know well. You can engage in an attention-demanding conversation while you drive and execute a complex sequence of cognitive and executive acts without conscious control.

Matthieu: The same is said about meditation: In the beginning, meditation is contrived and artificial, and gradually it becomes natural and effortless.

Wolf: As I briefly mentioned before, during the acquisition of skills, a shift occurs from cortical to subcortical systems. In the beginning, when conscious control and focused attention are required, neocortical structures have to be engaged, in particular those involved in attention in the frontal and parietal lobes. However, once the skill has been acquired and becomes more automatic, activity in cortical control systems decreases and other structures become more involved. In the case of motor skills, these structures are in addition to the cerebellum and the basal ganglia, the motor areas of the cortex that are always required.

Matthieu: It's clearly explained in meditation instructions that the first stages of meditation are always somehow artificial and require sustained efforts. Whether it comes easily in a spontaneous flow on some days or you feel bored on other days, you have to maintain the continuity and meditate day in and day out. It is said that it is better to do many short, regular, and repeated sessions of meditation rather than lengthy sessions only every week or fortnight.

Wolf: Which is exactly what you need to do when learning a skill learning and forming procedural memories. Much research has been done on the dynamics of skill learning and the neuronal substrate of procedural memory. It would be interesting to see whether the strategies that have been discovered as optimal for acquiring skills through procedural learning resemble those that have been intuitively worked out by teachers of meditation. Is it true, for example, that meditation sessions held just before you go to sleep are particularly effective? Because it is during sleep that procedural memory traces are shaped and consolidated.

CONSOLIDATING LEARNING THROUGH SLEEP

Matthieu: For most people who don't have the opportunity to do retreats or meditate a lot every day, it is said that the most important times are early in the morning and before going to sleep. By meditating or doing any type of spiritual practice early in the morning, you set the tone for the day and set in motion a process of inner transformation that will somehow carry on through the day's activities, like an invisible stream. To use another image, the fragrance of the meditation will remain and give a particular perfume to the whole day. It will create a different atmosphere, a different attitude, a way of being, and a way of relating to your own emotions and those of other people. Whatever happens during the day, you have an inner state of mind to which you can return. During the day, from time to time, you can also rekindle the meditation state, even for short moments, to enhance your experience.

Before falling asleep, if you clearly generate a positive state of mind, filled with compassion or altruism, it is said that this will give a different quality to the whole night. Oppositely, if you go to sleep while harboring anger or jealousy, then you will carry it through the night and poison your sleep. This is why a practitioner will endeavor to maintain a positive attitude right up to the moment of falling asleep, trying to maintain a clear and luminous state of mind. If you do so, then the flow will continue through the night.

Wolf: This theory corresponds well to recent data on the relevance of sleep for learning and memory processes. It is well established by now that you have to go through a repetitive sequence of characteristic sleep patterns to consolidate memories—slow-wave sleep, so-called deep sleep, and the paradoxical sleep, rapid eye movement (REM) sleep, during which the brain is highly active and exhibits electrographic patterns indistinguishable from being awake, aroused, and attentive. These sleep patterns alternate during the night and serve to reestablish the equilibrium of the brain. Because of its plasticity, the brain undergoes changes while it responds to the environment. Throughout the day, new memories are formed, new skills are acquired, and all this is associated with changes in myriad synaptic connections. To maintain stability, the networks have to be recalibrated in response to these changes, and this recalibration seems to occur during sleep. Memory traces become reorganized, the relevant is segregated from the irrelevant, and newly learned contents get embedded in their respective association fields.[31]

This is the reason that the contents of dreams are often related to events of the preceding day. The sleeping brain reactivates these memory traces to work on them, integrate them with previous traces, and consolidate them. During the early phases of sleep, the activity patterns caused by the experiences preceding sleep are replayed by the brain, often in time lapse (i.e., on a contracted time scale). This could be the explanation for why meditators report that the state they achieve right before going to sleep carries into their sleep.[32] However, this experience is not specific to meditation. Many people have experienced that learning the vocabulary of a foreign language is most efficient if one rehearses the list of words just before closing one's eyes. During sleep, the rehearsed contents are consolidated in the absence of interfering experiences and are usually retrievable with great clarity the next morning.

Matthieu: Also, when you need to make an important decision and feel a bit confused and uncertain about it, if you clearly put the question of what you should do in your mind before falling asleep, the next morning the first thought that comes into your head seems to indicate the most meaningful choice, the one that is less distorted and biased by mental projections, hopes, and fears.

Wolf: This is why we say *Schlaf darüber*, "Sleep on it," when a difficult problem needs to be solved. Insight often presents itself on awakening.

Matthieu: I also wanted to mention that it is quite striking that people who engage in long-term meditation retreats—in Tibetan Buddhism in particular, practitioners do retreats that last more than three years—require much less sleep. These meditators come from different backgrounds. Some are monks or nuns, some are highly educated in Buddhist philosophy, others are not, and many are lay practitioners. Of course, all have different temperaments. Nevertheless, in our retreat center in Nepal, for instance, after about a year of practice, almost everyone works their way down to just four hours of sleep. They typically sleep from 10 at night to 2 or 3 in the morning depending on the individual, and they do so without forcing themselves and without feeling any signs of sleep deprivation. They feel fresh during the day and don't doze off during their practice.

It is true that during the day, they don't face a great amount of novelty, they are not exposed to stressful circumstances, and they don't have to deal with all kinds of events and situations. Nevertheless, they are far from being inactive, and the schedule of meditation exercises is quite demanding. It includes intense cultivation of attention, compassion, visualization techniques, and other skills. How would you interpret such a striking physiological change?

Wolf: Here, several points come to my mind. First, it is known that young children (and this also holds for young animals), who have a lot to learn during the day because nearly everything they experience is novel to them, need to sleep much more than adults. They have to cope with massive changes in the functional architecture of their brains—and presumably not only because they have to learn more than adults but also because their brains are still developing, with new connections being formed and inappropriate connections removed. These massive circuitry modifications require permanent recalibration and therefore long sleep episodes. Actually a positive correlation exists between the amount of sleep you need and the amount of novelty you have to digest. If one enriches the experience of animals or humans during daytime, then it increases the length of sleep, and it also changes the sleep pattern. This correlation between the novelty of experiences (i.e., the amount of disequilibrium inflicted by learning-induced changes in the functional architecture of the brain), on the one hand, and the amount of sleep required, on the other hand, supports the notion that sleep is required to reinstall the brain's homeostasis.

Complex dynamic systems with a high degree of plasticity, such as the brain, are susceptible to disturbances of their equilibrium and are permanently at risk of entering critical states. When subjects are deprived of sleep for a long period of time, their brains become unstable. One consequence can be epileptic seizures, which are a reflection of supracritical excitatory states. In fact, doctors sometimes use sleep deprivation to diagnose epilepsy because it makes these pathological patterns more likely to occur. Other equally dramatic consequences of sleep deprivation are disturbances of cognitive functions with delusions, illusions, and even hallucinations. Memory functions deteriorate because there is no time to consolidate the engrams and regain equilibrium after the imbalance generated by learning. Moreover, attentional mechanisms deteriorate, which further impedes cognition and learning.

But let's come back to the question of why meditators in retreat need less sleep. I would think it's because they do not have to deal with novelty. They work on known and already stored contents, and thus there is probably less to be reorganized. The main task is consolidation, and this can perhaps be achieved even in the meditative state, as the brain is only minimally exposed to external stimulation.

Matthieu: That's right. The cultivation of skills and their consolidation is actually the main work of meditation. A study carried out in Madison, Wisconsin, in Giulio Tononi's laboratory, in collaboration with Antoine Lutz and Richard Davidson, showed that, among meditators who had completed between 2,000 and 10,000 hours of practice, the increase of gamma waves was maintained during deep sleep, with an intensity proportional to the number of hours previously devoted to meditation.[33] The fact that these changes persist in these people at rest *and* during sleep indicates a stable transformation of their habitual mental state, even in the absence of any specific effort, such as a meditation session.[34]

Wolf: It would be interesting to look at the sleep patterns people display while in retreat. The different sleep phases may have different functions, and it has been hypothesized that one serves consolidation while the other reestablishes equilibrium and orthogonalizes memory traces to reduce superposition and merging of memories that need to be kept segregated.

Matthieu: Can you explain that?

Wolf: The brain's memory is associative; it's not like computer memory, where you have distinct addresses for distinct contents. In the brain, different memories are stored within the same network by differential changes in the coupling of neurons. The equivalent of a particular engram is a specific dynamic state of the network, a state characterized by the specific spatiotemporal distribution of active and inactive neurons of the network.

Let me put it simply. Assume we have 26 interconnected neurons, A to Z, that can become active in different combinations because their connections have been strengthened and weakened in a specific way through previous learning. The memory trace of content 1 would then consist of the predisposition of neurons ACD to be active simultaneously, content 2 would correspond to the coactivation of neurons AMZ, and so on. Now, if you store more and more contents into this network, you run into problems of superposition. The neuronal representations of different contents may become too similar or merge with one another, such that the representations may become blurred and ambiguous.

Matthieu: Can you give an example?

Wolf: The boundaries between the assemblies of neurons representing different memories may become blurred because the same synapses, the same connections, may have to be used for the representation of different contents. The greater the number of different patterns you want to store in this network, the more sophisticated the arrangement of the coupling of the neurons must be to keep these patterns separate from each other and distinguishable.

Matthieu: Is it like when there are too many images in a mirror?

Wolf: Or too many different transparencies laid on top of each other. They will fuse and become blurred through interference. So you have to arrange your transparencies in a way that minimizes superposition and thus optimizes discriminability. Here is an example. You try to retrieve a name from memory, but its retrieval is blocked by another similar name that keeps popping up instead. In this case, the neuronal representations of the two names are not sufficiently segregated or orthogonalized. Improving the segregation of overlapping representations is thought to be one of the functions of sleep. One of the mechanisms could be to better bind the two names to their respective association fields, such as the differing contexts in which they have been stored. Whether the two functions, consolidation and orthogonalization of memory traces, are served by different sleep phases is still unknown, but it is conceivable that the one is more associated with consolidation and the other more with cleaning and arranging things properly. Therefore, it would be interesting to investigate the sleep patterns of meditators in retreat to see which phases dominate during their reduced sleep time.

Matthieu: Well, one indication might come from body movement. I was told that an average person turns over more than 15 times during a night. When His Holiness the Dalai Lama was told this fact by a sleep specialist, he wondered whether we really move that much. I was also a bit puzzled. Some meditators sleep through the night in a sitting position, cross-legged, when they do long retreats. So we would not expect them to move that much. Other meditators traditionally sleep on their right side, with the right hand resting under their cheek and the other one extended along their body.

Wolf: Is there a reason?

Matthieu: It's quite complex. The teachings say that by sleeping in this position, we press down and inhibit the subtle channels of the body that are on the right side, which are said to carry negative emotions, while we facilitate the movement of the energy through the left side channels, which carry positive emotions. It is intriguing that this fits well with the notion that the right prefrontal cortex is related with negative emotion, whereas the left one is activated when positive emotions are experienced. Another reason for sleeping on the right side is not to press down on the heart.

Some years ago, when I was doing an eight-month retreat, I tried to observe myself. Once or twice a night, I would look at the small clock on my table to check the time. Over seven months, I tried to notice my position when I awoke. Every time, whenever I woke up in the middle of the night or when I was about to get up in the morning, I just had to open my eyes to see the clock right there in front of my eyes. I never found myself even once gazing at the ceiling or facing the other way. So I am pretty confident that I didn't turn much during the five or so hours that I was sleeping.

Wolf: This would then suggest that you alternate much less between different sleep phases because these turns tend to occur during transitions.

Matthieu: While you go from a dream to deep sleep and so forth.

Wolf: Right, although nowadays people think that you also dream in slow-wave sleep. The structure of the dreams may be different, but the brain is working in both phases, and high-frequency oscillations also occur in the slow-wave sleep phases. These fast oscillations are superimposed on the slow waves, and because fast oscillations are likely to be associated with the recall or activation of memories, dreaming may also occur in these deep sleep phases. It's a bit difficult to find out because if you wake up people, you don't know whether what they tell you has been experienced right at that moment or they are remembering dreams from earlier in the night.

Matthieu: Is there a connection between remembering a dream more or less clearly between these two kinds of sleep?

Wolf: Again, it's difficult to say. I think the literature says that if you wake people up during the REM sleep period, the paradoxical sleep phase, then the probability of having a dream recalled is higher than if you wake them out of deep sleep. However, I don't know how valid the statistics are.

In the morning, you have a lot of paradoxical sleep phases. They increase throughout the night and reach a peak in the morning just before you wake up. Usually you remember the dreams that occur in the morning much more frequently than those that occur in the middle of the night—unless you have a dramatic dream that wakes you up. This finding would suggest that dreams occurring in REM sleep are more easily remembered.

COMPASSION AND ACTION

Matthieu: You spoke about beta and gamma waves and their function in creating some kind of synchrony or resonance among different parts of the brain, which makes you ready for a task. This process seems to be in agreement with meditation. It appears, for instance, that when meditators engage in compassion meditation, which activates gamma waves more powerfully than any other meditation practice, there is a particular activation of the frontal areas of the brain, the executive part, although the meditator is not doing anything. From the contemplative perspective, we could interpret that as a complete readiness to act for the sake of others, which is a natural quality of genuine altruism and compassion. If you are not caught inside the bubble of self-centeredness and are less involved in relating everything to yourself, then the ego ceases to feel threatened. You become less defensive, feel less fear, and are less obsessed with self-concern. As the deep feeling of insecurity goes away, the barriers that the ego created fall apart. You become more available to others and ready to engage in any action that could be benefit them. In a way, compassion has popped the ego bubble. That's our interpretation. That's why those states of compassion and open presence give the strongest gamma waves of all meditation states, more than focused attention, for instance.

Wolf: I think focused attention would engage much less neuronal substrate because you concentrate on a certain subtask, which in terms of brain architecture means you are using a certain subsystem. You activate and prepare this subsystem to come into play quickly by condensing all your resources in this subsystem.

Matthieu: The technical term in Tibetan for this meditation translates to "one-pointed focused attention."

Wolf: This should then lead to focused gamma activation of the areas that process the contents to which you are attending. One would expect the most widespread coherence to occur in association with the state that you call "complete openness"—or what is the right term?

Matthieu: Open presence. Of course, these words are approximate. It is quite difficult to put such experiences into words. But it turns out that unconditional compassion produces even higher gamma activation than open presence.

Antoine Lutz and Richard Davidson showed that when one plays recordings alternately of a woman crying out in distress and a baby laughing to experienced meditators in a state of compassion, several areas of the brain linked to empathy are activated, including the insula. This zone is more activated by the distress cries than by the baby's laughter. A close correlation is also observed among the subjective intensity of meditation on compassion, the activation of the insula, and cardiac rhythm.[35] This activation is all the more intense when the meditators have more hours of training. The amygdala and cingulate cortex are also activated, indicating increased sensitivity to others' emotional states.[36] Your daughter Tania and her team have also shown that the neural networks for altruistic compassion and empathy are not the same. Compassion and altruistic love have a warm, loving, and positive aspect that "stand-alone" empathy for the suffering of the other does not have. The latter can easily lead to empathic distress and burnout. While collaborating with Tania, we arrived at the idea that burnout was in fact a kind of "empathy fatigue" and not "compassion fatigue," as people often say.[37]

Barbara Fredrickson and her colleagues also showed that meditating on compassion for 30 minutes per day for six to eight weeks increased positive emotions and one's degree of satisfaction with existence.[38] The subjects felt more joy, kindness, gratitude, hope, and enthusiasm, and the longer their training was, the more marked were the positive effects.

COMPASSION, MEDITATION, AND BRAIN COHERENCE

Matthieu: These results of compassion meditation must represent a state of high coherence because the mind is entirely filled with benevolence and lovingkindness for all and compassion for those who suffer. To begin, you might focus on a particular object. To arouse lovingkindness, you will imagine, for instance, a lovely child and feel nothing but benevolence toward that child. When that mental state has clearly arisen in your mind, you make it grow and sustain it until it fills your whole mental landscape. Then you simply try nurturing this state, to keep it present, full, and vast.

Wolf: Let me try an interpretation of that. It may be wrong; it's speculative. The brain must be able to differentiate between good and bad states, consistent and inconsistent states.

Matthieu: In what way?

Wolf: The brain must know whether a particular dynamic state is a valid result, either of a perceptual act or a deliberation process, or whether the state is still part of the computations that eventually lead to the result. All you have in the brain is spiking neurons. There is always activity, a continuous stream of ever-changing activity patterns. The brain is constantly performing computations on incoming signals and searching for the most likely or plausible interpretations, and at a particular moment in time, a sudden transition occurs toward a result. We are aware of this transition, and therefore the brain must be able to distinguish activity that is supporting the computations toward a result and activity that represents the result.

Matthieu: Can you give an example of the result?

Wolf: There are many examples ranging from solutions to intellectual puzzles, riddles, and mathematical problems, to the apparently effortless solutions to perceptual problems. Imagine a cluttered scene with figures on a complex background. Your visual system performs complicated computations to isolate the figures from the background and identify them through comparison with stored knowledge, and all of a sudden there is this moment—Eureka! I have it! I have recognized it, I have the solution. This solution is a particular spatiotemporal activation pattern that is not so different from the patterns generated while the brain is searching for the solution. The question is how the activation patterns, which represent a solution, differ from those that reflect the computational process that leads to the solution. There must be a signature to neuronal states that are solutions that does not vary regardless of the content of the solution. In addition, this signature must have various degrees of magnitude because we are to ascribe various degrees of reliability to a particular solution.

Matthieu: In what way?

Wolf: Some solutions are not very trustworthy. Perhaps I get to the end of a deliberation process but—

Matthieu: —you are not satisfied.

Wolf: Right. I realize that this is not a reliable solution or that it is only a preliminary solution. I have to continue, or I have no solution at all, and I have to do more computation. The brain must have evaluation systems that are able to distinguish these activation states. Otherwise you wouldn't know when to stop a computation and talk about the result, and you wouldn't know what the quality of the result is. Therefore, the brain must have a way to evaluate internal states: "This is a satisfying state. This is a dissatisfying state." These value-attributing systems also support learning processes because you want to favor states that are identified as good states, and you want to disfavor states identified as bad states. Thus, the polarity of the changes in synaptic connectivity that mediate the learning has to be switched as a function of the value attributed to a state. That is, to favor the reoccurrence of a state, the connections between neurons supporting that state should be strengthened and vice versa: Connections favoring adverse states should be weakened to render generation of such states less likely.

Matthieu: This is similar to the fundamentals of training happiness as a skill, first recognizing which emotions and mental states are disturbing and which favor happiness. Then one works at fading out the former and developing the latter.

Wolf: Let me come back to your statement that the training and experience of compassion is also a pleasant state.

Matthieu: I would rather call it a fulfilling state.

Wolf: As the electroencephalographic data indicate, this state is a highly coherent state. It is associated with a high degree of synchrony of high-frequency oscillations. Now comes my speculation. Maybe the signature of a solution is coherence, a state of synchrony, the moment at which ensembles of neurons engage in well-synchronized oscillatory activity. All the value-assigning systems would have to do to detect such coherent states is take samples of the actual activity patterns present in cortical networks and determine the degree of coherence or measure the amount of synchronicity. This task is not difficult because neurons are able to distinguish between synchronous and temporally dispersed input, the former being much more efficient in driving the neurons. As long as the patterns are temporally dispersed and rapidly changing, the value-assigning systems would not be activated, and the cortical activity would be classified as resulting from ongoing computations that have not yet converged toward a result. If, however, the activity has become highly coherent, the value-assigning systems would become active, signaling that a result has been obtained.

Now, if the signature of a result is the coherence of a state, the transient synchronization of a sufficient number of neurons distributed over a sufficiently large number of cortical areas that lasts sufficiently long to be considered valid or stable, the internal evaluation centers would signal that a result has been obtained and enable learning mechanisms to fix that state in case it needs to be remembered, strengthening those connections that support this particular state.

Let me now extend this: We know from experience that it is pleasant to arrive at results. "Eureka" can be an extremely fulfilling feeling. Thus, activation of evaluation centers appears to be associated with positive emotions—one of the reasons that we sometimes work hard to obtain solutions. Maybe we have an explanation here for the rewarding emotions associated with meditation. As the available data suggest, you generate an internal state during meditation that is characterized by a high degree of coherence, by the synchronization of oscillatory neuronal activity across an extended network of cortical areas. This condition should be ideal for the activation of the evaluation systems that detect globally coherent states and reward solutions with positive feelings. What you practice in meditation is perhaps the generation of such globally coherent states without, however, focusing on particular contents. You generate the state that has all the signatures of a good and reliable solution without any specific content. Extrapolating from how it feels to have obtained a solution to a concrete problem, I imagine that you get a feeling of content-free harmony, a feeling that all conflicts are resolved and everything has fallen into place.

Matthieu: Yes, we call it fulfillment, wholesomeness, inner peace. That brings us back to what His Holiness the Dalai Lama often says, with a good touch of humor, when he explains that the *bodhisattva*—the ideal embodiment of altruism and compassion in the Buddhist path—has in fact found the smartest way to fulfill his own wish for happiness. The Dalai Lama adds that when thinking and acting in an altruistic way, it is not at all guaranteed that we will actually benefit others or even please them. When you try to help someone, even with a perfectly pure motivation, they might look at you suspiciously and ask, "Hey, what do you want, what's the matter with you?" But you are 100% sure to be helping yourself because altruism is the most positive of all mental states. So the Dalai Lama concludes, "The *bodhisattva* is smartly selfish." In contrast, the one who only thinks of himself is "foolishly selfish" because he only brings distress on himself.

Wolf: It seems indispensable that the brain can evaluate its own states, distinguish the unwanted from the wanted, and then attach emotions to these states so that unwanted states can be avoided and wanted states approached. If you have an inner conflict, this state is unwanted, and you are driven by unpleasant feelings to get out of that state, to search for a solution. So there must also be a characteristic signature to activation patterns that represent a conflict. This signature should again have a generalizable format because, although conflicts can have many different reasons and arise at many different levels of processing, the feelings associated with conflicts are similar, suggesting a final common path for the evaluation of conflicts. However, to the best of my knowledge, we do not yet know what the neuronal signature of conflict is. It must again be a particular state of activity—maybe a particularly low level of coherence.

Matthieu: Many people are literally destroyed by inner conflicts.

Wolf: We know from animal research that certain reward systems modulate their activity as a function of conflict, and evidence indicates that, among other regions, the anterior cingulate cortex monitors internal conflicts.[39]

Matthieu: Inner conflicts go together with a lot of rumination.

Wolf: Yes, but we ignore the nature of the activity patterns representing conflicts and ruminations. Maybe it is a condition where mutually exclusive assemblies compete for prevalence, thereby causing instability, a permanent alternation between metastable states—

Matthieu: —we simply call that "hope and fear"—

Wolf: —if no stable state is reachable, if the internal model of the world that the brain permanently has to update by learning continues to be in disagreement with "reality." If the brain is striving for stable, coherent states because they represent results and can be used as the basis for future actions, and if pleasant feelings are associated with these consistent states, the one purpose of mental training could be to generate such states in the absence of any practical goals. However, to generate such states right away, detached from any concrete content, may be difficult. This is probably the reason that you initially imagine concrete objects—why you try to focus attention on specific, action-related emotions to evoke positive feelings such as generosity, altruism, and compassion, which are all highly rewarding attitudes.

Matthieu: As opposed to selfish behavior.

Wolf: Exactly. So you use this imagery as a vehicle to generate coherent brain states, and if the contents are pleasant, then a joyful condition is created. Then, once you gain more expertise in controlling brain states, you learn to detach these states from their triggers until they become increasingly free of content and autonomous.

ALTRUISM AND WELL-BEING

Matthieu: I think it may be a bit unfair to view compassion as merely a pleasant experience because compassion and fulfillment are highly intertwined. Human qualities often come in a cluster. Altruism, inner peace, strength and freedom, and genuine happiness grow together like the various parts of a nourishing fruit. Selfishness, animosity, and fear come together as the parts of a poisonous plant. The best way to become truly compassionate is out of wisdom, by deeply realizing that others do not want to suffer, just like you, and want to be happy, just as you do. Consequently, you become genuinely concerned with their happiness and suffering. Helping others may sometimes not be "pleasant," in the sense that you might have to deliberately endure some "unpleasant" hardship to help someone, but deep within is found a sense of inner peace and courage and a sense of harmony with the interdependence of all things and beings.

Wolf: You are right. If you are able to combine your own well-being with altruism and compassion, it is a win-win situation. However, what I am asking is, what state does the brain identify as a pleasant state? We know that we have those states, when we have a solution, when we have resolved a conflict, when we have helped others—particular brain states must also be experienced as good states. As the electrophysiological correlates of your meditation states suggest, these good states are apparently states of high coherence among many areas of the cerebral cortex. Is this at all a plausible assumption?

Matthieu: You know better than me about coherent states, but it makes sense to me. To come back to inner conflicts, they are mostly linked with excessive rumination on the past and anxious anticipation of the future, and thus they lead to being tormented by hope and fear.

Wolf: I see it as an exaggeration of the otherwise well-adapted and necessary attempt to use past experience to predict the future, an attempt that is likely to not always converge toward a stable solution because the future is not foreseeable. Maybe it is the clinging to the fruitless search for the best possible solution—that is by definition impossible to find—that frustrates the system and causes the uneasy feelings.

MAGIC MOMENTS

Matthieu: "I am anxious to know whether this will happen or not." "What should I do?" "Why do people behave like that with me?" "I am so worried that they people are saying things about me." Such streams of thought lead to unstable states of mind. That feeling of insecurity is reinforced by hiding in the bubble of self-centeredness to protect oneself. In fact, within the confined space of self-centeredness, rumination goes wild. Thus, one of the purposes of meditation is to break the bubble of ego grasping and let these mental constructs vanish into the open space of freedom.

When in daily life people experience moments of grace or magic, when walking in the snow under the stars or spending a beautiful moment with dear friends by the seaside or on top of a mountain, what happens? All of a sudden, the burden of inner conflict is lifted. They feel in harmony with others, with themselves, with the world. They feel good, and so the inner conflict disappears for some time. It is great to fully enjoy such magical moments, but it is also revealing to understand why they felt so good: pacification of inner conflicts and a better sense of interdependence with everything, rather than fragmenting reality into solid, autonomous entities, a respite from mental toxins. All of these are qualities that can be cultivated through developing wisdom and inner freedom. This practice will lead not to just a few moments of grace but to a lasting state of well-being that we may call genuine happiness. It is a satisfactory state because the feelings of insecurity gradually give way to a deep confidence.

Wolf: Confidence in what?

Matthieu: Confidence that you will be able to use those skills to deal with the ups and downs of life, sensations, emotions, and so on in a much more optimal way. Your equanimity, which is not indifference, will spare you from being swayed back and forth like mountain grass in the winds by every possible blame and praise, gain and loss, comfort and discomfort, and so on. You can always relate to the depth of inner peace, and the waves at the surface will not appear as threatening as before.

COULD FEEDBACK REPLACE MIND TRAINING?

Wolf: Thus, through mental training, you familiarize yourself with states of inner stability, thereby protecting yourself against fruitless ruminations. If these desirable states have a characteristic electrographic signature that can be measured and monitored, then we could use biofeedback to facilitate the learning process required to obtain and maintain these states. It might help to familiarize oneself with these states more quickly. Admittedly, this approach is a typical Western aspiration to circumvent cumbersome and time-consuming procedures and look for shortcuts on the way to happiness...

Matthieu: You know about the experiments during which scientists implanted electrodes in a region of the brain in rats that produces sensations of pleasure when stimulated. The rats can stimulate themselves by pressing a bar. The pleasure is so intense that they soon abandon all other activities, including food and sex. The pursuit of this feeling becomes an insatiable hunger, an uncontrollable need, and the rats press the bar until they die from exhaustion. So, I am convinced that any shortcut will result more in a state of addiction than in a deep change in your way of being, as is acquired through mind training. The perception of inner peace and fulfillment is a byproduct of having developed an entire cluster of human qualities. Grasping onto perpetually renewed pleasant sensations would probably achieve different results. As for feedback, however, as you know, a research project is underway with your daughter Tania which seems to indicate that expert meditators, when given feedback about the activation state of particular brain areas, can modulate at will compassion, attention, and even negative feelings such as disgust or intense physical pain.[40]

Wolf: But in this case, you first generate the characteristic activation patterns associated with these emotions and learn something about the quantitative relation between activity and the intensity of a feeling. My question was whether you consider it possible to enhance by trial and error certain brain states that are displayed to you via a feedback loop and then, step by step, get more and more familiar with these states until you can generate them intentionally.

Matthieu: It may not be impossible, but I don't think it's the best way to proceed, and I don't really see the point of doing so. Simply getting feedback on one particular skill may not help beginners because all of these skills—attention, emotional control, empathy—need to be developed simultaneously, and that is what meditation techniques do. Also, the continuous, long-term use of these techniques, what we call *meditation*, is based on wisdom, on a deep understanding of the way the mind works, and on the nature of reality (as being impermanent, interdependent, etc.). Feedback techniques may lack the richness of contemplative methods used to develop empathy, altruism, and emotional balance. However, for therapeutic purposes, such feedback training might have great virtues and help people who lack a particular skill, such as attention or empathy, to focus more on developing that skill. They also might help us understand better how the brain functions.

Further, mere feedback or, even worse, direct stimulation may not result in changes in ethical behavior, as in the case of meditation. Stimulating areas of the brain that induce nice sensations, for instance, or taking drugs that make you feel high all the time is unlikely to make you become a more compassionate and ethical human being. It might even leave you feeling more dependent and powerless than ever once the stimulation has ceased.

Wolf: Some drugs seem to directly activate the structures in the brain that normally would become active only as a consequence of brain states that we addressed as the "good" states—the states free of conflict, the states corresponding to solutions, the coherent states. The drugs obviously do not generate these complex states but act directly on the systems that evaluate these states. They fool these value-assigning systems.

Matthieu: That is why generating one pleasant sensation after another just to feel good would be at best an impoverished version of mind training and could even have opposite effects. It is impoverished in the sense that the entire idea of the wholesome flourishing of a human being comes from cultivating a vast array of qualities ranging from wisdom to compassion that are meant to achieve genuine happiness and a good heart, which can be said to be the goal of life.

Wolf: Since the flower power era, psychoactive drugs have often been advocated as a means to open doors toward a better understanding of oneself, by widening one's realm of experience and creating altered states of consciousness that can be remembered and even cultivated once the drug's effects have faded. Since around the same time, biofeedback has been propagated as a technique to enter states of relaxation. If you are in a relaxed, idling state, with eyes closed, large regions of the brain engage in synchronized oscillatory activity in the alpha-frequency range around 10 Hz. This activity can be measured easily. If its amplitude is converted into a tone and subjects are asked to try to increase the intensity of that tone, then one observes after some time that subjects are indeed able to increase their alpha activity—and they also report entering states of relaxation.

Matthieu: Interestingly enough, a preliminary study with meditators trained in Tibetan Buddhism indicated that when one suddenly clears the mind of all mental chatter, alpha waves all but disappear for a while. After the detonation that normally triggers a startle response, the meditator's mind is left in a crystal-clear state devoid of mental constructs and discursive thinking. So, if these results are confirmed, then a meditator would relate alpha waves to mental chatter, the little chaotic conversation that seems to be going on most of the time in the background of our mind.

Wolf: Certainly alpha is suppressed when you focus your attention on something. It is reduced when you open your eyes and engage in scrutinizing the environment, an observation which suggests that the alpha a rhythm is incompatible with the engagement of attention. It appears to serve the suppression of potentially distracting activity. However, other evidence suggests that alpha serves as the carrier rhythm for the coordination of higher frequency oscillations and thereby might play a role in the attention-dependent formation of functional networks. I would predict that there is little alpha when you are engaged in your type of meditation because you produce this massive gamma activity, and high gamma activity usually excludes alpha activity.

Matthieu: As mentioned earlier, we should correct the naïve image of meditation that still predominates in the West as sitting somewhere to empty your mind and relax. Of course, an element of relaxation is present, in the sense of getting rid of inner conflict, cultivating inner peace, and freeing oneself from tensions. Also, an element of emptying your mind can be seen, in the sense of not perpetuating mental fabrications or linear thinking and resting in a state of the clear freshness of the present moment. However, this state is neither "blank" nor dull relaxation. It is a much richer state of vivid awareness. Also, one does not try to prevent the thoughts from arising, which is not possible, but frees them as they arise.

Wolf: When you say "linear thinking," you mean serial thinking, going from one item to the next, the typical feature of conscious deliberation.

Matthieu: I mean discursive thinking, chain reactions of thoughts and emotions, which results in constant mental noise.

Wolf: Rumination, then, is endless spirals of thoughts.

Matthieu: Yes, it's a nonstop proliferation of thoughts based on self-centeredness, hopes, and fears that are mostly afflictive.

Wolf: It would be interesting to perform a thorough study of the electrographic signature of this chattering or mind wandering. It would perhaps be feasible if it were possible to deliberately switch between chattering and nonchattering states. It would be nice to know the signature of these states of serial processing that make you go from A to B to C in search of a solution. It should be the signature of the states that precede decisions.

Matthieu: When people are lost in their thoughts, that automatic process may go on for quite some time, and suddenly they find themselves somewhere else completely.

Wolf: That can be agreeable. It sometimes happens when you read—at least that happens to me. You read and then discover that your eyes continue to go through the lines, but your mind is somewhere else, following a different path. This experience is not necessarily unpleasant, and on occasion it may even spark a creative moment. It may, all of a sudden, provide you with new insights, with unexpected solutions.

Matthieu: That could be, if you evoke wholesome memories or situations, but still it pertains to the wandering mind, which creates an obstacle to clarity and stability. If you want to evoke particular states of mind, then it would be best to do it with a sense of direction, which contributes to inner flourishing, rather than just drifting along with your thoughts.

Wolf: Unstable states need not be unpleasant, and I suspect they could even be an indispensable prerequisite for creativity.

Matthieu: Research done by the neuroscientist Scott Barry Kaufman has indicated that brain states favorable to creativity seem to be mutually exclusive with focused attention. According to him, creativity is born from a fusion of seemingly contradictory mental states that can be limpid and messy, wise and crazy, exhilarating and painful, spontaneous and yet arising from sustained training.[41]

Wolf: I wish to come back to discussing the virtues of aversive brain states. As mentioned already, there must be brain states that are identified as desirable and are associated with positive feelings. Likewise, there must be states that the brain tries to avoid—that are associated with or lead to aversive feelings. The latter are certainly as important as the former in guiding our behavior, promoting learning, and protecting us from running into dangerous traps, just as pain is as important as satisfaction. The aversive states motivate the search for solutions that resolve the conflict. Now, aren't there two complementary strategies to avoid the aversive feelings associated with conflicting brain states? One would consist of taking drugs that dampen the systems causing the aversive feelings—it is like taking painkillers to avoid pain. Along the same lines, you can repress the bad feeling using distraction or attending to something else or you learn to tolerate the aversive state. These strategies allow you to cope with the symptoms but don't address the cause.

Or, and this is the second strategy, you try to go to the root, find out what the problem is, and then try to resolve it. This strategy may require rumination and inflict temporarily even more aversive feelings, but it may eventually lead to the elimination of conflicting causes and converge to a solution. Both strategies have their virtues. Clearly, the elimination of causes is always preferable. However, if the conflict is not resolvable, if the costs for its resolution are too high, or if it is an illusory conflict, drugs, repression, or learning to cope may be more appropriate than striving for a final but unattainable solution. To which of the two strategies do you think meditation comes closest? Could it be that it is a technique to resolve conflicts by simply dissociating the aversive feelings from the problem, by dissolving the feeling and not coping with the problem itself? If this were the case, then I would assume it is a practice ensuring well-being only in a highly protected environment like a monastery or some other similarly ideal circumstances.

Then there is a third state that I would like to call the idling state, which is associated with neither good nor bad feelings. The brain hovers above ground, ready to engage but not yet engaged in any goal-directed activity.

Matthieu: We call that a *neutral* or an *undetermined state*. It is neither positive nor negative, but it is still imbued with mental confusion.

Wolf: Every human being is familiar with these three states—positive, neutral, and aversive—because our brains are built that way. We have all developed strategies to avoid the negative states and to stay as long as possible in the former two—and I assume that evolution programmed us that way to ensure our survival in a complex, uncertain world. Now you claim that with mental training, we can reprogram our brains so that we can spend more time in the positive states, with the dual effect that we feel better and at the same time behave more in a way that reduces conflicts in our interactions with the world. This would indeed be the key to heaven on earth, and to me it sounds too good to be true. Where then are the limits of this seemingly golden way to global happiness?

ARE THERE LIMITS TO MIND TRAINING?

Matthieu: There are obvious limits to our physical capacities, such as how fast we can run or how high we can jump, and to most of our mental faculties, such as the number of items we can store in our short-term memory, our ability to remain perfectly attentive to a task for a long time, the amount of information we can process simultaneously when faced with many different stimuli, and so forth, although such faculties can be trained to an extent. However, can there be a limit to human qualities, such as compassion or lovingkindness, which are more like qualia than quantities? I don't see why, no matter what level of compassion is now in our mind, we cannot continue to increase it, to make it even clearer and vaster. I don't see what could limit that. You can always conceive of a fuller, deeper, more intense compassion. This is an important point for mind training. His Holiness the Dalai Lama often says that, yes, there are limitations to what one can learn in terms of information, but compassion can be developed boundlessly.

Wolf: Interesting. I would have thought the reverse. Compassion is an emotional state, so its amplitude should somehow be encoded in the activity of neurons mediating this state. Learning leads to engrams, and because of the combinatorial nature of engrams that we discussed earlier, the storage capacity of the brain is enormous. This is exemplified by the well-known case of certain idiot savants, who have difficulties in conceptualizing abstract relations and therefore have to use the brute force of memory to cope with the world. The vast number of neurons and the even more numerous connections modified by learning allow the brain to produce a virtually infinite number of different engrams. If a particular memory were implemented by a particular activation pattern in which something like 100 neurons participate at any one moment in time, then you have more possible states, at least mathematically, than you have atoms in the universe.

Matthieu: One hundred neurons connecting with every other neuron?

Wolf: No, if you take the 10^{11} neurons but you define a state as a specific constellation of 100 neurons, then of course you have many different 100-neuron packages in this space of 10^{11} neurons. Each of them can have many different states, and this gives you this unimaginably large number of possible states. This incredibly high-dimensional storage space is, however, not fully exploitable because of the superposition problem mentioned earlier. Engrams must not overlap too much to remain retrievable as separate representations.

Matthieu: Then how come we don't all have such incredible faculties to memorize and calculate like the autistic savants?

Wolf: In fact, we all have it but we don't realize it. If you take your visual memory, for example, you can come to a place somewhere in the world where you haven't been for 20 years and you will recognize that a house is missing and remember that you have been there before.

Matthieu: Or you come to a house where you have not been for a long time and immediately recognize tiny details such as the shape of a door knob or a teapot, when only a few minutes before arriving you could not have described them to save your life.

Wolf: We take this for granted because everyone can do this. Think about the memory capacity that is required to store all the scenes that you have perceived in your life. It is gigantic! You know how much memory is required to store a simple picture taken with a digital camera. We go through the world and store all the scenes that we perceive. We do not and cannot recall them at any time, but if the association cues are provided, we realize that they are stored quite reliably. So we all have this memory capacity, but it is not economical to use this brute force strategy of remembering everything to solve problems. It's much more economical to derive a rule and remember the rule, or just remember where you can find the information. It is more elegant to conceptualize and store a reduced, abstracted representation. Why store a complex situation with all the details if you can represent it with a single symbol, a single word?

Young children have less access to this economical strategy because they still have to acquire the ability to form abstract descriptions and use symbols. Therefore, they rely more on the brute storage strategy, and this is the reason that they always beat you in games like Memory. If the acquisition of the strategy relying on concepts and symbols is impaired because of a brain injury or genetic disposition, then the brute force strategy to remember everything in detail is cultivated. It helps to some extent to cope with life, but many of those autistic savants struggle with normal life because they have difficulties in representing complex relations.

Matthieu: And emotion?

Wolf: Some of the "memory giants" are actually deficient in recognizing emotions. Autistic children, for example, have difficulties in deciphering facial expressions and associating emotions with them, which makes it difficult for them to deduce from their caretakers' facial expressions whether what they are doing is right or wrong. This prevents them from attaching to their social environment and developing social skills. When children act, they always look at the caretaker at some stage to find out whether their behavior is appropriate. If they cannot interpret the caretaker's expression, it becomes difficult for them to develop cognitive functions and to attach to the world, and then they get more and more isolated. Here is another reason for the seemingly enormous memory capacity of certain autistic children:

Because they cannot invest in social relations, their environment gets impoverished, and they direct their interest toward other sources of information, such as timetables, calendars, and so on, and then they practice, rehearse, and memorize. That said, I am not a specialist in this field, and you should take everything I just said with caution.

Matthieu: I recently learned about someone, Daniel Tammet, who is able to recite, in proper order, 22,514 decimals of the number Pi. It took him more than five hours, and he did it without a single mistake. He said that he did not feel anxious about remembering such an incredible amount of numbers, only about reciting them in public. He also mentioned that, in general, whenever he feels anxious, just thinking of numbers calms him down and makes him feel secure.[42]

Wolf: These examples illustrate impressively that brains are capable of performances that go way beyond the imagination of most of us and that, at first sight and without proof of the contrary, we would have judged as impossible. Maybe the same holds for achievements that can be accomplished with intensive mental training. I would not know, however, how to quantify the magnitude of compassion because it is hard to measure phenomena that only exist in the first-person perspective.

The question of whether there is an upper limit to human qualities such as compassion and lovingkindness is not easy to resolve. Let's assume for a moment that these emotions are encoded in the intensity or salience of neuronal responses. In principle, there are three ways to increase the saliency of neuronal responses. One consists of increasing the discharge rate of the neurons. This strategy is applied for the encoding of sensory stimuli: The stronger the stimulus, the higher the discharge frequency of the encoding neurons. The second strategy is to recruit more and more neurons: The stronger the stimulus, the more neurons will respond because even the less excitable neurons will become active. The third strategy is to increase the synchronicity of the neuronal discharges because synchronous activity drives target neurons more effectively than temporally dispersed input, and therefore synchronous activity propagates more easily and more rapidly within neuronal networks.

Apparently, the brain can use the three strategies interchangeably, as we have recently been able to show in a study published in Neuron.[43] We exploited a visual phenomenon called *contrast enhancement*. The perceived contrast of a grating pattern can be enhanced if it is displayed on a background grating that differs from the target grating either in its orientation or by the relative spatial offset of the stripes. In both cases, the perceptual effect of enhanced contrast is exactly the same, but recordings from the neurons in the visual cortex that respond to the target show that in the first case neurons become more active, whereas in the second case their activity stays the same but becomes more synchronized. Still, there should be an upper limit. Once all neurons in the respective structure discharge at their maximal frequency and in perfect synchrony, no further increase in saliency is possible. So, I would predict that there is a limit to the intensity of an emotion.

Matthieu: I should add that when I said that one can always conceive of a fuller and deeper compassion, I did not refer only to the emotional aspect of compassion. Understanding the deeper causes of others' suffering and generating the determination to alleviate them also arises from wisdom and "cognitive" compassion. The latter is linked to the comprehension of the more fundamental cause of suffering, which, according to Buddhism, is ignorance—the delusion that distorts reality and gives rise to various mental obscurations and afflictive emotions such as hatred and compulsive desire. So this cognitive aspect of compassion can embrace the infinite number of sentient beings who suffer as a result of ignorance. I don't think we have to worry about the magnitude of such cognitive compassion exhausting the capacities of the brain.

MEDITATION AND ACTION

Matthieu: To come back to one of your initial questions, I should reply to the accusation of selfishness and indifference that is sometimes leveled at hermits and meditators. Such opinions reflect a deep misunderstanding about the Buddhist path because freeing oneself from the influence of self-centeredness and ego clinging is precisely what makes you more concerned with others and less indifferent toward the world. Meditation is a key process for developing and enhancing altruistic love and compassion. You could argue that it would be even better if the hermit left his hermitage to go and help people. Otherwise, what is he contributing to society? How would he learn about human interaction when remaining alone in his hermitage? This argument makes sense at first glance. Yet there are simple answers to these questions. You need time and concentration to cultivate a skill. While thrust into the often hectic conditions of the world, you might be too weak to become strong, too weak to help others and even to help yourself. You don't not have the energy, concentration, and time to train. So this developmental stage is necessary, even if it does not appear to be immediately useful to others.

When you build a hospital over a few months or years, the plumbing and electricity works do not cure anybody, yet when the hospital is ready, it provides a much more powerful tool to treat patients. It is worthwhile taking the time to build such a hospital, rather than just saying, "What's the point in waiting? Let's operate in the street." Hence, the idea is to develop skills in an environment that is conducive to mental training, so that one becomes strong enough to display and maintain genuine altruism and compassion even in the most trying and adverse circumstances, when it is most difficult to remain altruistic. I have now been exposed to the world of humanitarian activities for a number of years, and I have seen over and again that the main problems that plague the humanitarian world—corruption, clashes of ego, weak empathy, discouragement—stem from a lack of maturity in human qualities. So the advantages of spending dedicated time to develop human qualities are obvious. You thus gain inner strength, compassion, and balance *before* embarking on serving others.

Developing the right motivation is a crucial factor to everything we do. The famous Tibetan saint and poet Milarepa said that during the 12 years he spent in solitary retreats in the wilderness, there wasn't a single moment that wasn't dedicated to others. By this he meant that he was dedicated to developing the qualities needed to truly benefit others. In the Buddhist path, the core motivation of the apprentice *bodhisattva* is,

"May I achieve enlightenment in order to gain the capacity to free all beings from suffering." If such an aspiration is genuinely present in your mind, then your practice is the best investment you can make for the benefit of others. This is not the result of indifference but of the sound reasoning that you have to prepare yourself and build up the necessary strength to be of use to humanity.

Wolf: So it should become an integral part of life; you should go through this period in some stage of your development, but then you should interact with others again and not stay in the hermitage or in the protected environment of monasteries. For teachers it may be appropriate to stay in monasteries because they will transmit their wisdom to the pupils, but others should stay there only temporarily and then engage in the world for its betterment.

Matthieu: It also makes sense to see that as long as you are still a mess yourself, there's no point going out and messing around with other people's lives as well. You need to be skillful in recognizing when you are mature enough to meaningfully help others. Otherwise, it is like cutting the wheat when it is still green. Nobody benefits from it.

Wolf: But then, as long as you judge yourself not to be ready yet, you need to be supported by others who earn a living and have no chance to enjoy this protected maturation period. There are cultures in which these mental training practices do not exist at all, and yet somehow those ethnic groups have made it: They developed ethics, rules, morality; they reproduced; and they got along quite well. So it can't be the only way—it might be, but need not be, the best way, we don't know—but surely there must be other ways that also bring forth good people and stable societies.

Matthieu: Yes, of course, but somehow one must come to a point where one looks within and sincerely tries to become a better person.

Wolf: Is it then impossible to be raised and educated in a conventional way and still develop the ability to sharply distinguish your feelings, to feel empathic, to be a good parent, and so forth? From what we have discussed so far, there does not seem to be much chance for the development of all these good qualities if one is not lucky to become the pupil of a highly trained practitioner. Isn't education also an effective and perhaps complementary strategy to improve human qualities? It might even have the advantage that it does not require so much cognitive control as meditation and therefore can be applied early in life, when the brain is still developing, plastic, and shapeable. There is probably a tradeoff between the time that you invest into your own betterment and the time left to improve conditions in the world by interacting with it. The strategy I would prefer is to teach adolescents the practice of mental training so they can use it throughout life to get to know themselves better and become more equilibrated. Still, I would think the most effective way to modify brain functions to the better is education of the young by good caretakers. Children learn by assimilation from caretakers to which they have established tight emotional bonds. Thus, we might be able to bootstrap this process by having mental training as an obligatory subject in teaching curricula.

Matthieu: People who are gifted might be able to develop all these qualities in the midst of many other distracting activities. However, for most people, it does help to gather all one's strength from time to time to nurture human qualities with single-pointed dedication. It is quite possible that someone who naturally has a good heart would immediately succeed in helping others, more than a meditator who starts with a grumpy, selfish mind. The point is that they should both continue to improve themselves further, and cultivation does help vastly.

We should not underestimate the power of the transformation of the mind. We all have the potential for change, and it is such a pity when we neglect to actualize it. It is like coming home empty-handed from an island made of gold. Human life has immense value if we know how to use its relatively short time span to become a better person for one's own happiness and that of others. This requires some effort, but what doesn't? So let's end on a note of hope and encouragement: "Transform yourself to better transform the world."

Earlier, you also alluded to the periods between meditation sessions. This is a central topic in meditation praxis. We call them "postmeditation periods." These two phases, meditation and postmeditation, should reinforce each other. In fact, this idea was recently demonstrated in a real-life experiment by Paul Condon and Gaëlle Desbordes, who followed three groups for eight weeks. One group was trained in meditation on lovingkindness, one was trained in meditation on mindfulness, and the third, the control group, was left without any training. After eight weeks, the participants' altruistic behavior was put to the test by observing the probability that they would offer their seat in a waiting room to someone standing against the wall with crutches, showing signs of discomfort. Before the suffering individual enters, the participant is seated on a bench next to two other people (accomplices of the experimenter, along with the "sick" person) who don't show the least bit of interest in the standing patient (which accentuates the "bystander effect" that is known to inhibit helping behavior). Strikingly, on average, the meditators offered their seat *five times more often* than nonmeditators.[44]

So postmeditation activities and attitudes should reflect and express the qualities developed during meditation. There would be no point in achieving some fine meditation state of inner clarity and stability if you drop it completely the moment you stop meditating. Ideally, at some point, the meditator's skill and experience should be such that the two states begin to fuse.

Wolf: The same might apply for our discussion. Maybe we should rest and let our brains reorganize what we have learned and continue tomorrow.

DEALING WITH SUBCONSCIOUS PROCESSES AND EMOTIONS

ON THE NATURE OF THE UNCONSCIOUS

What is the unconscious? For the Buddhist monk, the most profound aspect of consciousness is alert presence. For him, what psychoanalysis calls the unconscious only represents the random mists of mental fabrications. For the neuroscientist, precise criteria distinguish between conscious and unconscious processes, and it is important to identify everything that happens in the mind as it prepares for conscious cognitive processes. Then comes the question of emotions. How to neutralize conflicting situations? How does altruistic love differ from passionate love? Is love the highest emotion? Points of view coincide concerning the effectiveness of cognitive therapy.

Matthieu: Let's explore for a bit the notion of the unconscious, from neuroscientific and contemplative perspectives. Usually when people speak about the unconscious, they refer to something deep in our psyche that we cannot access with our ordinary consciousness. We certainly have the concept, in Buddhism, of habitual tendencies that are opaque to our awareness. These tendencies initiate various thought patterns that can either occur spontaneously or be triggered by some kind of external circumstance. Sometimes you are just sitting there, thinking of nothing in particular, and suddenly the thought of someone or a particular event or situation pops up in the mind, seemingly out of nowhere. From there, a whole chain of thoughts begins to unfold, and if you are not mindful, you can easily get lost in it.

The general public, psychologists, and neuroscientists surely have varying views about what the unconscious is. As for psychoanalysis, what it calls the depths of the unconscious are, from a contemplative perspective, the outer layers of clouds formed by mental confusion that temporarily prevent one from experiencing the most fundamental nature of mind. How can there be something unconscious in a state of pure awareness, devoid of mental construct? No darkness exists in the middle of the sun. For Buddhism, the deepest, most fundamental aspect of consciousness is this sun-like awareness, not the murky unconscious. Of course, this is all expressed from the first-person perspective, and I am sure that a neuroscientist approaching this issue from the third-person perspective will have a different view of the unconscious.

Wolf: Yes, I see it a bit differently. As mentioned earlier, an enormous amount of knowledge is stored in the specific architectures of the brain, but we are not aware of most of these "given" heuristics, assumptions, concepts, and so on. These routines determine the outcomes of cognitive processes, which we are aware of, but the routines remain hidden in the unconscious. Usually we are not aware of the rules that govern the interpretation of sensory signals, the construction of our percepts, or the logic according to which we learn, decide, associate, and act.

We cannot move these implicit hypotheses and rules to the workspace of consciousness by focusing our attention on them, as is possible with contents stored, for example, in declarative memory, the memory in which we store what has been consciously experienced. Abundant evidence indicates that attentional mechanisms play a crucial role in controlling access to consciousness. When attended to, most signals from our senses can reach the level of conscious awareness. Exceptions are certain odors, such as pheromones, that are processed by special subsystems and cannot be perceived consciously. Then there are the many signals from within the body that are excluded from conscious processing, such as messages about blood pressure, sugar level, and so on. It cannot be emphasized enough, however, that signals permanently excluded from conscious processing as well as transitorily excluded signals such as nonattended sensory stimuli still have a massive impact on behavior. In addition, these unconscious signals can control attentional mechanisms and thereby determine which of the stored memories or sensory signals will be attended and transferred to the level of conscious processing.

Another constraint is the limited capacity of the workspace of consciousness. At any one moment in time, only a limited number of contents can be processed consciously. Whether these limitations are due to the inability to attend to large numbers of items simultaneously or whether they result from the restricted capacity of working memory or both is still a matter of scientific investigation. The capacity of the workspace is limited to four to seven different items. This finding corresponds to the number of contents that can be kept simultaneously in working memory. The phenomenon of *change blindness*, the inability to detect local changes in two images presented in quick succession, demonstrates impressively our inability to attend to and consciously process all features of an image simultaneously.

Perception is actually not as holistic as it appears to be. We scan complex scenes serially, and actually much of what we seem to perceive we are in fact reconstructing from memory. Which of the many signals actually reach the level of conscious awareness is determined by a host of factors, both conscious and unconscious. It depends on what we attend to, and this is controlled by either external cues, such as the saliency of a stimulus, or internal motifs, many of which we may not actually be aware of. Then it may occur that even an attentive, conscious search for content stored in declarative memory fails to raise it to the level of awareness. We are all familiar with the temporary inability to remember an episode or a name and then how a persisting subconscious search process may suddenly lift the content into the workspace of consciousness. It appears that we are not always capable of controlling which contents enter consciousness.

I consider the workspace of consciousness as the highest and most integrated level of brain function. Access to this workspace is privileged and controlled by attention. Moreover, the rules governing conscious deliberations such as consciously made decisions most likely differ from those of subconscious processes. The former are based mainly on rational, logical, or syntactic rules, and the search for solutions is essentially a serial process. Arguments and facts are scrutinized one by one and possible outcomes investigated. Hence, conscious processing takes time. Subconscious mechanisms seem to rely more on parallel processing, whereby a large number of neuronal assemblies, each of which represents a particular solution, enter into competition with one another. Then a "winner-takes-all" algorithm leads to the stabilization of the assembly that best fits the actual context of distributed activity patterns. Thus, the conscious mechanism is suited best to circumstances in which no time pressure exists, when not too many variables have to be considered, and when the variables are defined with sufficient precision to be subjected to rational analysis. The domains of subconscious processing are situations requiring fast responses or conditions where large numbers of underdetermined variables have to be considered simultaneously and weighed against variables that have no or only limited access to conscious processing, such as the wealth of implicit knowledge and heuristics, vague feelings, and hidden motives or drives.

The outcome of such subconscious processes manifests itself in either immediate behavioral responses or what are called "gut feelings." It is often not possible to indicate with a rational argument why exactly one has responded in that way and why one feels that something is wrong or right. In experimental settings, one can even demonstrate that the rational arguments given for or against a particular response do not always correspond to the "real" causes. For complex problems with numerous entangled variables, it often turns out that subconscious processes lead to better solutions than conscious deliberations because of the wealth of heuristics exploitable by subconscious processing. Given the large amount of information and implicit knowledge to which consciousness has no or only sporadic access and the crucial importance of subconscious heuristics for decision making and the guidance of behavior, training oneself to ignore the voices of the subconscious would not be a helpful, well-adapted strategy.

Matthieu: What you said corresponds with what Daniel Kahneman explains in his book *Thinking, Fast and Slow*.¹ Although we are generally convinced that we are rational, our decisions, economic or otherwise, are often irrational and strongly influenced by our immediate gut feelings, emotions, and situations to which we have been exposed immediately before taking a decision. Intuition is a highly adaptable faculty that allows us to make fast decisions in complex situations, but it also lures us into thinking that we have made a rational choice, which actually takes more time and deliberation.

I understand that a lot is going on in the brain to allow us to function and have coherent perceptions, memories, and so on. But I was thinking more of the pragmatic aspect of dealing with the particular tendencies that give rise to the afflictive mental states and emotions associated with suffering. My point was that if you know how to relate to pure awareness and rest within that space of awareness, when disturbing emotions arise, they dissolve as they appear and do not create suffering. If one is an expert in this, then there is no need to bother about what is going on down in the subconscious. It is more a question of method. Psychoanalysis, for instance, contends that you need to find a way to dig into those hidden impulses and identify them, whereas Buddhist meditation teaches you to free the thoughts as they arise.

By dwelling in the clarity of the present moment, you are free from all ruminations, upsetting emotions, frustrations, and other inner conflicts. If you learn to deal, moment after moment, with the arising of thoughts, then you can preserve your inner freedom, which is the desired goal of such training.

SIDE EFFECTS OF MEDITATION

Wolf: Here we seem to face rather divergent concepts of the virtues of subconscious processing and the way we should deal with the unconscious dimension of our mind. This brings me to a critical issue: the side effects of meditation. One could argue that a strategy consisting of closing one's eyes when facing conflicts, in escaping from problems rather than solving them, is perhaps a suboptimal strategy. Let us assume that conflicts exist in the subconscious and that the rumination motivated by these conflicts serves to identify and settle them. Such conflicts could arise from ambiguous bonding between the child and her caretaker in early infancy or from conflicting imperatives imprinted by early education. The causes of such problems cannot easily surface in consciousness because they are part of implicit memories that have been formed prior to the maturation of declarative memory. Such conflicts jeopardize mental and physical health.

Humankind throughout its history has sought relief from such problems, with drugs, cultural activities, and, more recently, a host of specially designed therapies. Most of the latter require one to face the problem to cope with it. Another strategy, which is applied in cognitive and behavioral therapy, attempts to alleviate the problem by unlearning the habit using conditioning paradigms. If one suffers from a particular phobia, then one gets exposed to the threatening events and learns that they don't cause harm, and after a while, one habituates to the threat and the problem may be solved.

Matthieu: We surely need to get at the heart of the issue. But in the end, what we need is to be free from inner conflicts, one way or another, right? So, there might be ways that involve digging into the past as much as one can, with or without the help of a therapist, and then trying to solve the problem or trauma that has thus been identified, thereby freeing oneself from the afflictive effect. But there are also ways, including those used by Buddhism, that do not attempt to elude the problem but to free any conflicting thoughts that arise in the mind at the moment they arise. If you become expert in those methods, the so-called afflictive thoughts no longer have the power to afflict you because they undo themselves the moment they arise. But that is not all: Experience shows that by repeatedly doing so, you not only deal successfully with each individual arising of afflictive thoughts but you also slowly erode the tendencies for such thoughts to arise. So in the end, you are free of them entirely. Among contemporary Western therapies, cognitive and behavioral therapy also offer methods to attend precisely to a particular emotion that upsets you in the moment and deal with it in a reasonable and constructive way and has therefore some interesting similarities with the Buddhist approach.

Wolf: Let us see what meditation could contribute to the resolution of conflicts that arise at levels inaccessible to conscious processing. I shall take a critical stance and use a real-world problem as an example. Imagine a conflict evolves between two partners that evokes uneasy feelings and causes lasting mental rumination in both parties. Their two ego bubbles fight against each other, as in *Who's Afraid of Virginia Woolf?* Love and passion exist between them, which are both difficult to control because they are anchored in subconscious spheres. The partners go into retreat and meditate, stop ruminating, and feel fine while they meditate alone in a protected environment. But will this solve the problem? Will they not resume fighting once they are back home, together, and confronted again with their problems?

Matthieu: To openly confront our differences can be a way to pacify a conflict, but it is not the only one. To begin with, a conflict requires two protagonists confronting each other in antagonistic ways. One cannot clap with one hand only. In fact, if one of the persons involved disarms his or her own antagonistic mind, then it will contribute greatly to reducing the conflict with the other person.

We did an experiment at Berkeley with Paul Ekman and Robert Levenson, who, among other things, have been studying conflict resolution. In this case, they wanted me to have two conversations with two different people. The idea was to discuss a controversial topic—in this case, why a former biologist like me, who did research in a prestigious lab at Pasteur Institute, would ever choose not only to become a Buddhist monk but to believe in crazy things such as reincarnation. We were all fitted with all kinds of sensors detecting our heart rate, blood pressure, breathing, skin conductivity and sweating, and body movements, and our facial expressions were recorded on a video camera, to be later analyzed in details for fleeting microexpressions. My first interlocutor was Professor Donald Glaser, a Nobel laureate in physics, who then moved to research in neurobiology. He was an extremely kind and open-minded person. Our discussion went well, and at the end of the 10 minutes, we both regretted not having more time to dialogue. Our physiological parameters indicated a calm, nonconflictual attitude. Then came in someone who had to be chosen because he was reputed to be a rather difficult person—he was not told about that of course! He knew that we were supposed to get into a heated debate and went straight into it. His physiology became immediately highly aroused. From my side, I tried my best to remain calm—I actually enjoyed it—and did my best to provide reasonable answers delivered in a friendly way. Soon enough, his physiology became calmer and calmer, and at the end of the 10 minutes, he told the researchers, "I can't fight with this guy. He says reasonable things and smiles all the time." So, as the Tibetan saying goes, "One cannot clap with one hand."

As far as your own inner conflicts are concerned, if you use meditation simply as a quick fix to superficially appease your emotions, you temporarily enjoy a pleasant deferral of these inner conflicts. But as you rightly say, these cosmetic changes have not reached the root of the problem.

Merely putting problems to sleep for a while or trying to forcibly suppress strong emotions will not help either. You are just keeping a time bomb ticking somewhere in a corner of your mind.

True meditation, however, is not just taking a break. It is not simply closing one's eyes to the problem for a while. Meditation goes to the root of the problem. You need to become aware of the destructive aspect of compulsive attachment and all of the conflictive mental states that you mentioned. They are destructive in the sense of undermining your happiness and that of others, and to counteract them you need more than just a calming pill. Meditation practice offers many kinds of antidotes.

A direct antidote is a state of mind that is diametrically opposed to the afflictive emotion you want to overcome, such as heat and cold. Benevolence, for instance, is the direct opposite of malevolence because you cannot wish simultaneously to benefit and harm the same person. Using this kind of antidote neutralizes the negative emotions that afflict us.

Let's take the example of desire. Everyone would agree that desire is natural and plays an essential role in helping us to realize our aspirations. But desire in itself is neither helpful nor harmful. Everything depends on what kind of influence it has over us. It is capable of both providing inspiration in our life and poisoning it. It can encourage us to act in a way that is constructive for ourselves and others, but it can also bring about intense pain. The latter occurs when desire is associated with grasping and craving. It then causes us to become addicted to the causes of suffering. In that case, it is a source of unhappiness, and there is no advantage in continuing to be ruled by it. Here you may apply the antidote of inner freedom to the desire that causes suffering. You bring to your mind the comforting and soothing quality of inner freedom and spend a few moments allowing a feeling of freedom to be born and grow in you.

Because desire also tends to distort reality and project its object as something that you cannot live without, to regain a more accurate view of things, you may take the time to examine all aspects of the object of your desire and see how your mind has superimposed its own projections onto it. Finally, you let your mind relax into the state of awareness, free from hope and fear, and appreciate the freshness of the present moment, which acts like a balm to soothe the burning of desire. If you do that repeatedly and perseveringly—and this point is really the most important—this will gradually lead to a real change in the way you experience things all the time.

The second, even more powerful way to deal with afflictive emotions is to stop identifying with them: You are not the desire, you are not the conflict, and you are not the anger. Usually we identify with our emotions completely. When we become overwhelmed by desire, anxiety, or a fit of anger, we become one with it. It is omnipresent in our mind, leaving no room for other mental states such as inner peace, patience, or reasoning, which might calm our torments.

The antidote is to *be aware* of desire or anger, instead of identifying with it. Then the part of our mind that is aware of anger is not angry, it is simply aware. In other words, awareness is not affected by the emotion it is observing. Understanding that makes it possible to step back and realize that the emotion is actually devoid of solidity. We just need to provide an open space of inner freedom, and the internal affliction will dissolve by itself.

By doing so, we avoid two extremes, each as inefficient as the other: repressing our emotion, which would then remain as powerful as before, or letting the emotion flare up, at the expense of those around us and of our own inner peace. Not to identify with emotions is a fundamental antidote that is applicable to all kinds of emotions in any circumstance.

This method might seem difficult at the beginning, especially in the heat of the moment, but with practice, it will become easier to retain mastery of your mind and deal with the conflicting emotions of day-to-day life.

LOVE VERSUS ATTACHMENT

Wolf: And love, that almighty power that is at the origin of so much suffering and joy, would this wonderful force also fade and turn us into lukewarm beings without passion?

Matthieu: Not at all. The constructive aspect of love, at least altruistic love and what psychologist Barbara Fredrickson calls the "positive resonance" between persons, has no reason to disappear. In fact, when you are free from mental distortions, it becomes even stronger, vaster, and more fulfilling.

As for romantic love, there is usually a strong component of grasping and self-centeredness that will most often turn into a cause of torment. In this kind of love, one is often loving oneself through the guise of loving someone else. To be a source of mutual happiness, genuine love has to be altruistic. This does not mean at all that one will not flourish oneself. Altruistic love is win-win, whereas selfish love soon turns into a lose-lose situation.

Wolf: Can you selectively remove the grasping component?

Matthieu: Yes, because the grasping component is most often experienced as conflict and torment and the letting go of it as a relief. Joy is felt in letting go. Grasping means, "I love you if you love me the way I want." This situation is uncomfortable. How can you demand that someone should be the way you want her or him to be? That is unreasonable and unfair. Altruistic love and compassion can apply to any one person and can also be extended to all.

Wolf: How object centered is this?

Matthieu: The universal nature of extended altruism does not mean that it becomes a vague, abstract feeling, disconnected from reality. It should be applied spontaneously and pragmatically to every being who presents him or herself in the field of our attention. It can also focus more intensely on particular persons who are naturally close to us. The sun shines over all people equally, with the same brightness and warmth in every direction. Yet people in our lives—our family, our friends—who happened to be closer to the sun of our care, love, and compassion naturally receive more light and heat. This does not mean that this sun of compassion focuses its rays *only on them* in a discriminatory way, at the risk of actually shine less well on all.

Wolf: How does this differ from the dreams of the flower power movement—dreams that did not pass the test of reality? They were based on similar assumptions: just share love and compassion, and everybody will be happy. It didn't work out. Adolescents went from their natal family directly into the next family, the commune, without having been alone in between, without the chance to detach and mature. But it was seen as a way to get rid of the self-centered ego by sharing love, affection, responsibilities, and material goods. Is this an idea Buddhist societies would subscribe to?

Matthieu: I can't say what the flower power people really had in mind. When I mention extending love to more and more people, I did not mean being more and more promiscuous, of course. I meant extending altruistic love and compassion, which is quite different, isn't it? As far as Buddhism is concerned, extending altruistic love to all certainly does not come at the cost of neglecting one's own children. The example of the sun is apt: You give your full, undiminished love to those who are close to you, those for whom you are responsible, but you also preserve a complete openness and readiness to extend that altruism to whoever crosses your path in life. This does not have much to do with a collapse of personal relationships and with favoring sexual promiscuity, which is more likely to engender confusion and grasping. Unconditional altruism is a state of benevolence for all sentient beings, a state of mind in which hatred has no place.

Wolf: It is a characteristic feature of love that it hurts when the beloved other is not around. Seeking one's own pleasure may become sad if one cannot share that experience with one's beloved.

Matthieu: It should not only be when we see some beautiful natural scenery that we wish others could be there too; it should also be when we experience deep inner joy and serenity that we wish others may experience the same. This brings us back to the aspiration of the bodhisattva: "May I transform myself and achieve enlightenment so that I become able to free all beings from suffering." This is a much vaster and deeper aspiration than simply wishing that a loved one might be there to see a beautiful sunset with you.

Wolf: This sounds reasonable, generous, and mature, of course. But is it within the reach of human capacity? Human beings are uniquely able to bond with others, but when these bonds are stretched, they suffer. It seems to be a deep-rooted trait. When I listen to you, it seems to me as if you advocate a practice that promotes some kind of detachment. This can certainly reduce suffering and negative emotions such as hatred, revenge, envy, greed, jealousy, and aggression, but doesn't it also damage the amplitude of the other strong, precious, and utterly joyful feelings associated with highs and lows at the same time? Do you really believe that we can extend our compassion far beyond our personal relations? After all, our cognitive abilities and emotional capacities have been selected by evolution to cope with social interactions within small groups of individuals who know each other. Are you not taking equanimity in the place of intensity?

ON THE JOY OF INNER PEACE

Matthieu: I don't think so. Having inner peace and equanimity does not mean that you cease to experience things with depth and brilliance, nor does it necessitate a reduction in the quality of your love, affection, vivid openness to others, or joy. In fact, you can be all the more present to others and to the world because you are remaining in the freshness of the present moment instead of being carried away by wandering thoughts. What you call the "highs and lows" are like the surface of the ocean: sometimes stormy, sometimes calm. The effects of these highs and lows are heightened if you are near the shore, where there is little depth: sometimes you surf with euphoria on top of the wave, and the next moment you hit the sand or rocks and are in pain. But at high sea, when you have several thousand meters of depth below the surface, whether there are enormous waves or the surface of the ocean is like a mirror, the depth of the ocean below always exists. You will still experience joys and sorrows, but they will occur in the context of a much deeper, vaster mind. You may also remember that research in psychology and neuroscience indicates that states such as unconditional compassion and altruistic love seem to be the most positive among all positive emotions. Among various meditative states, the meditation on compassion is the one that produces the strongest activation of all. Researchers in positive psychology, such as Barbara Fredrickson, have concluded that love is the "supreme emotion" because, more than any other mental state, it opens our minds and allows us to view situations with a vaster perspective, be more receptive to others, and adopt flexible and creative attitudes and behavior.[2] It causes an upward spiral of constructive mental states. It also makes us more resilient, allowing us to manage adversity better. These states are anything but dull or indifferent.

Harmonious relationships can provide the most wonderful opportunity to enhance the reciprocal feeling of lovingkindness. But you have to build up the inner depth so that this love is vaster than the state of being enraptured by someone and by your own attachment to the feeling this love gives you. Evolution gave us this capacity to feel love for a person who is special to us. This is the case in parental love, particularly maternal love. But we can use this capacity as the foundation for extending the circle of love further and further.

Earlier you asked about difficulty in a personal relationship. Obviously that can bring intense pain. Yet if you overflow with love toward all living beings, the distress caused by the sudden loss of a dear one will be less disruptive because a vast amount of love still resides in your heart ready to be expressed toward many others.

If such distress occurs, then you should examine its nature. Is it provoked by the fact that your self-centered love has been upset? Does it prevent you from giving love to and receiving love from others? In truth, the more inner peace and contentment you experience, the more you can stand on your own two feet in a loving rather than an egotistical way.

Wolf: This reduced vulnerability to suffering is then not restricted to monastic life?

Matthieu: That would be a pretty limited application of inner strength! We all have the potential for it because it results from a genuine understanding of the ways the mind works and the cultivation of compassion and inner contentment.

Wolf: Perhaps we should impose two years of meditation practice before people get married, rather than military service.

Matthieu: Great idea! People should do that before engaging in any path in life, in fact. In the humanitarian world where I work, you can see that what often derail great humanitarian projects are human shortcomings: corruption, clashes of egos, and so on. The best training for nongovernmental organizations might be to ask all humanitarian workers to do a three-month retreat on altruistic love and compassion. Paul Ekman once told me that we should have a "compassion gymnasium" in every city. Truly, many ways exist to actualize the potential we have to enhance our own basic human qualities through mind training.

One of my teachers told me that to be able to feel unconditional compassion, one needs to develop fearlessness. If you are excessively self-centered, then you naturally feel insecure and threatened by everything around you. But if you are primarily concerned with others and not obsessed with yourself, why should you be so fearful? These qualities could be taught at school in a secular way as part of a program for cultivating emotional balance and fortitude. But for this we would need teachers who are familiar with the way emotions work.

Wolf: Many of us obviously suffer from exactly this problem. We feel uneasy because of some inner conflict, and then we go to work and divert our attention by concentrating on other immediate problems, repress our emotions, and get along until the problems inevitably find a way in again through the backdoor and then require even more effort to be masked and refuted. Obviously, this vicious circle could be interrupted if we invested time and effort from the outset to identify the nature of the shadows.

WATCHING THE MIND, TRAINING THE MIND

Matthieu: I think the cognitive therapy technique comes quite close to that. When I met him, Aaron Beck, the founder of cognitive and behavioral therapy, told me he was struck by its convergences with the approach of Buddhism.[3]

Figuring among the similarities Aaron Beck noticed was eliminating the "six main mental afflictions": attachment, anger and hostility, arrogance and mental confusion, which are to be slowly replaced by serenity, compassion and inner freedom. He also noted similarities in the application of procedures and meditation techniques aiming to reduce the mental fabrications leading to these afflictions: in particular, being absorbed in intransigent egocentricity.

One of the aims of cognitive therapies is in fact to gain awareness of the mental fabrications and exaggerations that individuals superimpose on certain events and situations. Cognitive therapies and Buddhism also aim to successfully reduce people's tendency to assign the highest priority, and sometimes exclusive priority, to their own objectives and desires, to the detriment of other people and often their own well-being and mental health. Beck notes that people suffering from psychotic problems experience intensified self-focalization: They relate everything to themselves and are exclusively concerned with the fulfillment of their own wants and needs. It must also be said that "normal" people often display the same type of egocentricity but to a lesser extent and in a more subtle way. Buddhism tries to diminish these characteristics.

What we really need is to identify the mental events that arise in our mind and skillfully resolve them. Many of the things that continue upsetting us are superimpositions on reality, mental fabrications that we can easily deconstruct. We need to be more skillful in paying attention to all the nuances of what is actually happening in our mind and in successfully freeing ourselves from being enslaved by our own thoughts. This is how one can gain inner freedom.

We invest a lot of effort in improving the external conditions of our lives, but in the end, it is always the mind that experiences the world and translates these outer conditions into either well-being or suffering. If we are able to transform the way we perceive things, then we will transform at the same time the quality of our lives. The normal mind is often confused, agitated, rebellious, and subject to innumerable automatic patterns. The goal is not to shut down the mind and make it like a vegetable, but rather to make it free, lucid, and balanced.

Wolf: I fully agree but wish to maintain that there may be many different ways to get there. Different cultures propose different strategies, extending from Socratic dialogues about the essence of things and the human condition to a host of spiritual practices, most of them embedded in religious systems, to humanistic stances based on the ideals of enlightenment, to distinct educational programs, and finally to therapeutic interventions. Would it not be desirable to devote some effort to the identification of the most efficient and practical strategies?

Matthieu: Of course there are many ways to get there. In Buddhism alone, one speaks of 84,000 gates to the path of liberation. What matters is which method actually works for you, according to your own mental dispositions, life circumstances, and capacities. If you want to open a door, then you need the right key. There is no point in choosing a golden key if it is the old rusted iron key that actually opens the door.

Wolf: As far as I can tell, the most efficient strategy so far has been the codification of human rights in modern, democratic constitutions and the installation of sanctions for violating social norms—hand in hand with the development of political and economic systems that protect the freedom of individuals and optimize equality. Surely these measures have to go together with transformations of the individuals embedded in these systems to have the greatest impact. If external causes for suffering are reduced and if social and governmental structures are such that compassion, altruism, justice, and responsibility are recognized and rewarded, then human beings are more likely to exhibit such traits and vice versa.

Matthieu: Society and its institutions influence and condition individuals, but individuals can in turn make society and institutions evolve as well. As this interaction continues over the course of generations, culture and individuals keep on shaping each other. Cultural evolution can be applied to both moral values—certain values are more apt to be transmitted from one individual to another—and beliefs in general, insofar as certain beliefs give people greater chances of surviving.

If we want to encourage a more caring and humane society to develop, it is important to evaluate the respective capacities for change of both individuals and society. If humans had no ability to evolve by themselves, it would be better to concentrate all our efforts on transforming institutions and society and not waste time encouraging individual transformation. But the experience of contemplatives, on the one hand, and the research on neuroplasticity and epigenetics, on the other, has shown that individuals can change.

Wolf: Of course. Otherwise we would not invest much hope in the possibility for education to nurture character traits that reduce individual suffering and contribute to the stabilization of peaceful societies. But the effectiveness of education is undisputed, and so is the value of attitudes that you claim result from contemplative training. It should be investigated whether current scientific evidence shows that meditation has the transformative power you advocate or that societies in which contemplative practices are widespread live in greater peace and inflict less suffering on their members than societies in which such practices are uncommon.

HOW DO WE KNOW WHAT WE KNOW?

WHAT REALITY DO WE PERCEIVE?

Can we understand reality as it is? On the level of ordinary perception, the neuroscientist and the Buddhist thinker say no: We never stop interpreting sensorial signals and constructing "our" reality. What are the advantages and disadvantages of this interpretation? Is it possible, with experimental and intellectual investigation, to shed light on the true nature of things? How do we acquire knowledge? Is there an objective reality independent from our perceptions? The *first-person* approach will be distinguished from *second- and third-person* exterior approaches. Is it possible to perfect our inner microscope, with introspection, and correct our distortions of reality and remedy the causes of suffering?

Wolf: Both of our traditions are deeply involved in fascinating epistemic questions: questions of how we acquire knowledge about the world, how reliable this knowledge is, and whether our perceptions reflect reality as it is or whether we perceive only the results of interpretations. Is it at all possible to recognize the "true" nature of the things around us, or do we only have access to their appearance? We have two different sources of knowledge to call on. The primary and most important source is our subjective experience because it results from introspection or our interactions with the world around us. The second source is science, which attempts to understand the world and our condition by extending our senses with instruments, applying the tools of rational reasoning to interpreting observed phenomena, developing predictive models, and verifying our predictions through experiments. Both sources of knowledge, however, are limited by the cognitive abilities of our brains because these constrain what and how we perceive, imagine, and reason. Precisely because of these constraints, we don't know where the limits of our cognition are; we can only posit that such limits are likely to exist.

Let me give a few examples to illustrate why I think that this is an inescapable conclusion. The brain is the product of an evolutionary process, just as every other organ and the human organism are as a whole. The brain is also the product of an undirected evolutionary game, in which the generation of diversity and selection has brought forth organisms optimized for survival and reproduction. As a consequence, these organisms are adapted to the world in which they have evolved. Life developed only in a narrow segment of the world as it is known to us: the mesoscopic range. The smallest organisms capable of autonomously sustaining their structural integrity and reproducing themselves consist of assemblies of interacting molecules that are confined by a membrane and measure only a few micrometers. One example is bacteria.

Multicellular organisms, plants, and animals reach sizes in the range of meters. All of these organisms have developed sensors for signals that are relevant for their survival and reproduction. Accordingly, these sensors respond to only a narrow range of the available signals. The algorithms developed for the evaluation of the registered signals have adapted to the specific needs of the respective organisms. Thus, the cognitive functions of organisms are highly idiosyncratic and tuned to a limited range of dimensions. For we humans, this is the world that we can perceive directly with our five senses and that we tend to equate with the "ordinary world." This is the dimension in which the laws of classical physics prevail, which is probably the reason that these laws were discovered before those of quantum physics. This is the segment of the world for which our nervous system generates well-adapted behavior, our senses define perceptual categories, and our reasoning leads to plausible and useful interpretations of the nature of objects and the laws that govern their interactions.

It follows from these considerations that our cognitive systems have, with all likelihood, not been optimized to unravel the true nature of perceived phenomena in the Kantian sense. Emanuel Kant distinguished between a hypothetical *Ding an sich*—literally the "thing in itself," or the essence of an object of cognition that cannot be reduced further to anything else—and the phenomenological appearance of that object, which is accessible to our senses. Our sensory organs and the neuronal structures that evaluate their signals have evolved to capture the information that is relevant for survival and reproduction and to generate behavioral responses according to pragmatic heuristics that serve these functions. Objectivity of perception (i.e., the ability to recognize the hypothetical *Ding an sich*) has never been a selection criterion. We know today that we only perceive a narrow spectrum of the physical and chemical properties of this world. We use those few signals to construct our perceptions, and our naïve intuition is that these provide us with a complete and coherent view of the world. We trust our cognitive faculties; we experience our perceptions as reflecting reality and cannot feel otherwise. In other words, our primary perceptions, whether mediated by introspection or sensory experience, appear to us as evident. They have the status of convictions.

Matthieu: We believe that we experience reality as it is, without realizing how much we interpret and distort it. Indeed, a gap exists between the way things appear and the way they are.

Wolf: Right. We have many examples illustrating this selective adaptation of our cognition to phenomena that matter for our life. One is our inability to develop an intuition about the phenomena that quantum physics postulates. The conditions of this microcosmos are difficult for us to imagine. The same holds true for the dimensions of the universe and the highly nonlinear dynamics of complex systems. Neither our sensors nor our cognitive functions have been adapted by evolution to cope with these aspects of the world because they were irrelevant for survival at the time when our cognition evolved.

Matthieu: For example, it is quite difficult to imagine something that appears as either a wave, which is not localized, or a particle, which is localized depending on the way you look at it.

Wolf: We also have difficulties with the cosmological scale. Take relativity theory: Our preconceptions resist the idea that the coordinates of space and time should be intertwined and relative because in the mesoscopic world, our ordinary world, we experience space and time as different and separate dimensions. It is interesting, though, that we are nevertheless able to explore dimensions of the world that are accessible to neither introspection nor our primary experience by extending our sense organs with instruments—telescopes and microscopes—and by complementing our cognitive abilities with the analytical and inductive power of reason. We make inferences, derive predictions, and validate them by experiment. However, all this occurs within the closed system of scientific reasoning, and there is no guarantee that the obtained insights have the status of irrefutability.

Matthieu: But here you are referring only to an understanding of reality based on our ordinary sense perceptions. When Buddhism speaks of apprehending "reality as it is," it does not refer to mere perceptions but to a logical assessment of the ultimate nature of reality. If you ask whether reality is made of a collection of autonomous, self-existing entities, and then conduct a proper, logical investigation, then you will come to the conclusion that what appear to exist as truly separate entities are in fact a set of interdependent phenomena devoid of autonomous intrinsic existence. However, even if one understands this mentally, that does not mean one's senses will perceive outer phenomena as they are, without any distortion. In fact, the Buddha himself said:

Eyes, ears, and nose are not valid cognizers.

Likewise the tongue and the body are not valid cognizers.

If these sense faculties were valid cognizers,

What could the sublime path do for anyone?[1]

Here, the "sublime path" refers to a proper investigation of the ultimate nature of reality.

Wolf: Before commenting on these deep insights obtained by contemplation, I would like to make a few additional remarks on the evolution of our cognitive systems, especially with respect to the transition between biological and cultural evolution. Our basic cognitive functions were initially selected to help us cope with the conditions of a presocial world. At the late stages of biological evolution, there was with all likelihood some coevolution between the emerging social environment and our brains, a coevolution that endowed our brains with certain social skills, such as the ability to perceive, emit, and interpret social signals. These abilities were then further complemented and refined by epigenetic modifications of brain architectures that occur during the development of individuals and are guided by experience and education.

Matthieu: *Epigenetics* refers to the fact that we inherited a set of genes, but the expression of these genes can be modulated by influences that we encounter during our lifetime. These can be outer influences, such as receiving a lot of affection or being abused, or inner influences, such as enduring severe anxiety or having a mind at peace. Some of these modifications can even occur in utero. When pregnant rats are exposed to chronic stress, their children have altered stress responses and react more sensitively to stress. The reason is that some of the genes that code for proteins involved in stress-regulating networks become downregulated, meaning they become less active and produce fewer or none of the proteins they are coding for. Recent research has shown that meditation has a significant effect on the expression of a number of genes, including some related to stress.[2]

Wolf: Yes, these additional modifications of brain functions are caused by imprinting and learning processes and serve as the major transmission mechanisms for sociocultural evolution. Thus, our brains are the product of both biological and cultural evolution and exist in these two dimensions. Through cultural evolution, those realities that we address as immaterial entities, as psychological, mental, and spiritual phenomena, came into being. These phenomena came into the world because of the special cognitive skills of human beings, which allow us to create social realities such as belief and value systems, and to conceptualize what we observe in ourselves and others, such as feelings, emotions, convictions, and attitudes. If all these phenomena are constructs of our brains—which is, to me, the most likely assumption—then their ontological status, their relation to "reality," should be subject to the same epistemic limitations that constrain our brains when it comes to perceiving the deeper nature of the world. Thus, the possibility needs to be considered that not only our perceptions, motivations, and behavioral responses but also our way of reasoning and drawing inferences are adapted to the particular conditions of the world in which we evolved, including the world of social realities that emerged during cultural evolution.

HOW DO WE ACQUIRE KNOWLEDGE?

Matthieu: Let's recall that epistemology is the theory of knowledge, the philosophical discipline that investigates the methods used to acquire knowledge and distinguish valid cognition from mere opinions and naïve perceptions.

Wolf: From a neurobiological perspective, the distinction between "valid cognition" and "naïve perceptions" is not evident. We consider perception as an active, constructive process, whereby the brain uses its a priori knowledge about the world to interpret the signals provided by the sense organs. Remember, brains harbor a huge amount of knowledge about the world. Unlike computers—which possess separate components for the storage of programs and data and for the execution of computations—in the brain, all these functions are implemented in and determined by the functional architecture of the neuronal network. By "functional architecture," I mean the way in which neurons are connected to each other, which particular neurons are actually connected, whether these connections are excitatory or inhibitory, and whether they are strong or weak. When a brain learns something new, a change in its functional architecture occurs: Certain connections are strengthened, whereas others are weakened. Hence, all the knowledge a brain has at its disposal, as well as the programs according to which this knowledge is used to interpret sensory signals and structure behavioral responses, resides in the specific layout of its functional architecture. The search for sources of knowledge is thus reducible to the identification of factors that specify and modify the functional architecture of brains.

This leads one to the identification of the three major sources of knowledge about the world. The first, and certainly not the least important, is evolution because a substantial part of the brain's functional architecture is determined by genes. This knowledge, which pertains primarily to the conditions of the precultural world and has been acquired through evolutionary adaptation, is stored in the genes and expressed in the functional architecture of a newborn's brain. This knowledge is implicit—we are not aware of having it because we were not around when it was acquired. Still, we use it to interpret the signals provided by our sense organs. Without this immense base of a priori knowledge, we would be unable to make sense of our perceptions because we would not know how to interpret sensory signals. This inborn knowledge is subsequently complemented by extensive epigenetic shaping of the brain's neuronal architecture, which adapts the developing brain to the actual conditions in which the individual lives.

The human brain develops many of its connections only after birth, and this process continues until ages 20 to 25. During this period, numerous new connections are formed and many of the existing connections removed; this making and breaking is guided by the neuronal activity. Because after birth neuronal activity is modulated by interactions with the environment, the development of brain architectures is thus determined by a host of epigenetic factors derived from the natural and social worlds.

A considerable part of this developmentally acquired knowledge also remains implicit because of the phenomenon of childhood amnesia. Children before age 4 have only a limited capacity to remember the context in which they have experienced and learned particular contents. The reason is that the brain centers required for these storage functions—we call them *episodic, biographical,* or *declarative memories*—have not matured yet. Thus, although young children learn efficiently and store contents in a robust way through structural modifications of their brain architecture, they often have no recollection of the source of this knowledge. Because of this apparent lack of causation, knowledge acquired in this way is implicit, just as evolutionarily acquired knowledge is, and often assumes the status of a conviction—that is, its truth is taken for granted.

Like innate knowledge, this acquired knowledge is used to shape cognitive processes and structure our perceptions. Yet we are not aware that what we perceive is actually the result of such a knowledge-based interpretation. This has far-reaching consequences: the genetic dispositions and, even more important, the epigenetic, culture-specific shaping of different brains introduce profound interindividual variability. Thus, it is not surprising that different persons, particularly those raised in different cultural environments, are likely to perceive the same reality differently. Because we are not aware of the fact that our perceptions are constructions, we are bound to take what we perceive as the only truth and do not question its objective status. This constructive nature of perception makes it difficult to distinguish between valid and naïve cognition.

Matthieu: Sure, but logical reasoning and rigorous investigation can allow us to unmask the game played by mental fabricationsThe study of the evolution of cultures is a new discipline that has led to remarkable advances over the last 30 years, particularly under the impetus of two American researchers, Robert Boyd and Peter Richerson. According to them, two evolutions occur in parallel: the slow evolution of genes and the relatively fast evolution of cultures, which allows psychological faculties to appear that could never have evolved under the influence of genes alone—hence the title of their book, *Not by Genes Alone*.[3]

Boyd and Richerson think of culture as a collection of ideas, knowledge, beliefs, values, abilities, and attitudes acquired through teaching, imitation, and every other kind of socially transmitted information.[4] Human transmission and cultural evolution are cumulative because each generation has at its start knowledge and technological experience acquired by previous generations.[5] Most human beings are inclined to conform to dominant attitudes, customs, and beliefs. The evolution of cultures favors the establishment of social institutions that define and reward respect for behavioral norms and punish nonconformity to ensure the harmony of communal life. Still, these norms are not fixed: Like cultures, they evolve with the acquisition of new knowledge.

Wolf: Let me give an example of how extensively our a priori knowledge (i.e., cognitive schemata) determines what and how we perceive. The perception of a dynamic process and the derived predictions depend on whether the observer assumes that it is a process with linear or nonlinear dynamics. A prime example of linear dynamics is a mechanical clock. Each swing of the pendulum will, through a cascade of transmission wheels, cause a precisely determined advance of the arms—and if it is a perfect clock that is free of friction or other disturbing influences, then it can be predicted for an unlimited period of time how the arms will move and where they will be at a particular moment in time. The dynamics of the clock follow a continuous and fully deterministic trajectory. The function determining this trajectory is a combination of pendulum swings and the conversion of those swings into a rotation of the arms.

An example of nonlinear dynamics is a pendulum that swings over a surface on which three magnets are fixed in a triangular arrangement. Let's assume the pendulum is free to swing in all planes. In this case, the movement of the pendulum is determined not only by the forces of gravity and kinetic energy but also by the overlapping attractions of the three magnetic fields. When one sets the pendulum in motion, it will move on extremely complex trajectories before it finally settles over one of the attracting magnets, but it is entirely unpredictable over which magnet it will come to rest. Even if the pendulum always starts from exactly the same position, it is not possible to predict where it will finally stop. Along the trajectory among the various attracting forces, there are many points where the pendulum could go right or left with equal probability; minute forces determine which way it finally goes. The pendulum's trajectory would also remain unpredictable if it were kept moving by driving it gently with an oscillating force. In that case, the pendulum would rotate around the three magnets. One could calculate the probability with which the pendulum would be, on average, within the attraction field of one of the magnets, but it would still be impossible to predict when exactly the pendulum would be where. If one were to control a clock with such a "chaotic" pendulum, then one would observe highly unpredictable movements of the arms.

For normal human perception, assume linearity is a well-adapted strategy for two reasons. First, many of the relevant processes around us can be approximated by linear models. Second, because the evolution or trajectories of linear systems are much more predictable than those of nonlinear systems, it provides little advantage to develop intuitions about the dynamics of a nonlinear system if one wishes to make inferences about the future dynamics of that system. For these reasons, there has probably been no selection pressure to evolve intuitions for highly nonlinear processes. As a consequence, we seem to have difficulty imagining processes that have nonlinear dynamics and drawing the right conclusions about these processes. For example, because we intuitively assume linearity, we misperceive the complex dynamics of economic or ecologic systems, nurture the illusion that we can forecast and hence control the future trajectories of these systems, and then are surprised when the outcome of our interventions differs radically from what we had expected. Given these evolutionary limitations of our cognitive abilities and intuitions, we are left with the burning question of which source of knowledge we should trust, especially when we are confronted with contradictions among our intuitions, primary perceptions, scientific statements, and collectively acquired social convictions.

Matthieu: Buddhism also emphasizes the fact that a correct understanding of the phenomenal world acknowledges the fact that all phenomena arise through almost numberless interdependent causes and conditions that interact outside of a linear causality.

CAN THERE BE VALID COGNITION OF SOME ASPECTS OF KNOWLEDGE?

Matthieu: According to Buddhist views, it is quite clear that a gap exists between the way things appear and the way they are. Can we close this gap and, if so, how?

When one is thirsty and sees a mirage in the desert, one may run toward it, hoping to get water, but of course the water is just an illusion. Many such examples show that the way things appear quite often doesn't correspond to reality. When the way we perceive outer phenomena is dysfunctional—we cannot drink the mirage water—in Tibetan Buddhism, this is called *invalid cognition*. If our mode of perception is functional, as when we perceive as water what we usually call water—something we can wash with and use to quench our thirst—it allows us to function in the world, and it is therefore called *valid cognition*. At a deeper level of analysis, which Buddhism calls *ultimate logic*, if we perceive water to be an autonomous, a truly existing phenomenon, this is considered to be an invalid cognition. Conversely, if we recognize that water is a transitory, interdependent phenomenon resulting from a myriad of causes and conditions, yet ultimately devoid of intrinsic reality, this becomes valid cognition.

At the level of perceptions, according to the Buddhist theory of perception, which has been debated by commentators for 2,000 years, at the initial moment of perception, our senses capture an object. Then comes a raw, nonconceptual mental image (of a form, sound, taste, smell, or touch). At a third stage, conceptual processes are set in motion: Memories and habitual patterns are superimposed on the mental image depending on the ways our consciousness has been shaped by past experiences. This gives birth to various concepts: We identify this mental image as being, for example, a flower. We superimpose judgments on it and interpret the flower as being beautiful or ugly. Next, we generate positive, negative, or neutral feelings about it, which in turn leads to attraction, repulsion, or indifference. By then, the phenomenon outside has already changed because of the transitory nature of all things.

So, in fact, the consciousness associated with sensory experiences never directly perceives reality as it is. What we perceive are images of past states of a phenomenon that are ultimately devoid of intrinsic properties. On a macroscopic level, we know, for instance, that when we look at a star, we are actually looking at what that star was many years ago because it has taken that many years for the light emitted by the star to reach our eyes. In fact, this is true of all perceptions. We are never looking directly at phenomena in real time, and we always distort them in some way.

What's more, the mental image of the flower (or any object) is also deceptive because we generally perceive the flower as being an autonomous entity and believe that the attributes of beauty or ugliness belong intrinsically to the flower. All of this proceeds from what Buddhism calls *ignorance* or *lack of awareness*. This kind of basic ignorance is not just a mere lack of information, such as not knowing the name of the flower, its therapeutic or poisonous effects, or the way it grows and reproduces. *Ignorance* here refers to a distorted and mistaken way of apprehending reality at a deeper level.

In essence, we should understand that what I perceive as being "my world" is a crystallization triggered by the encounter between my particular version of human consciousness and a vast array of outer phenomena not fully determined in and of themselves and that interact in nonlinear ways. When this encounter occurs, a particular perception of these outer phenomena occurs. Someone with insight will understand that the world we perceive is defined by a relational process taking place between the consciousness of the observer and a set of phenomena. It is therefore misleading to ascribe intrinsic properties to outer phenomena, such as beauty, ugliness, desirability, or repulsiveness. This insight has a therapeutic effect: It will disrupt the mechanism of compulsive attraction and repulsion that usually results in suffering.

But let's return to the initial question. Yes, it is possible to transcend deluded perception and achieve a valid understanding of the true nature of the flower as being impermanent and devoid of intrinsic, autonomous existence, as being devoid of any inherent qualities. Achieving this understanding is not dependent on our sensory perceptions or past habits. It comes from a proper analytical investigation of the nature of the phenomenal world, culminating in what is known in Buddhism as *all-discriminating wisdom*, an insight that apprehends the ultimate nature of phenomena without superimposing mental constructs on them.

Wolf: Can I just make a comment on how this would fit well with modern neuroscientific views? Evidence from psychophysical investigations of perception and neurophysiological studies on perceptions underlying neuronal processes suggests that perceiving is essentially reconstructing. The brain compares the sparse signals provided by our eclectic sense organs with the vast basis of knowledge about the world that is stored in its architecture and generates what appears to us as a percept of reality.

When we perceive the outer world, we first arrive at a coarse match between sensory signals and knowledge-based hypotheses about the world, and then we usually enter an iterative process to obtain approximations that gradually converge to the optimal solution—a state with a minimal number of unresolved ambiguities. We perform an active search for the best matches between signals and hypotheses until we obtain results with the desired clarity. This latter process of active search-and-match requires the investment of attentional resources, takes time, and is interpretative in nature. What is actually perceived is the result of that comparative process. It appears as if this scenario is fully compatible with your views! It suffices to replace what I address as "a priori knowledge" with what you term "consciousness."

Matthieu: There are two different ways of phrasing that: one from the third-person perspective, in the language of neuroscience, and the other from the first-person perspective, based on introspective experience. You described how our perception of the world is shaped by evolution and the increasing complexity of the nervous system. From a Buddhist perspective, one would say that our world, at least the world we perceive, is intimately intertwined with the way our consciousness functions. It is clear that, depending of their configuration and their past history, different streams of consciousness, whether human or not, will perceive the world in a different way. It is almost impossible for us to imagine what the world of an ant or a bat looks like. The only world we know results from the interdependent relation between our particular type of consciousness and the phenomenal world, which is a complex set of relations among countless interdependent events, causes, and conditions.

Just take the example of what we call an *ocean*. On a beautiful calm day the ocean appears to us like a mirror made of water, whereas on another day it might be the scene of a wild storm. We can relate these two conditions and still call these various states *ocean*. But what will a bat make out of the ultrasonic echoes sent up from a perfectly flat sea one day and from a chaotic sea filled with gigantic waves the next day? This is beyond our imagination in the same way that quantum physics eludes our ordinary representations. Thus, Buddhism says that our phenomenological world, the only one we perceive, depends on the particular configuration of the consciousness we have and is shaped by past experiences and habits.

Wolf: This is perfectly compatible with the Western perspective. The philosopher Thomas Nagel stated clearly that it is impossible for us to imagine how it would feel to be a bat.[6] The qualia of subjective experience are simply not translatable. Although we humans are endowed with a highly differentiated communication system, our language, and with the ability to imagine the mental processes of the respective human "other," it is still difficult—if not impossible—to know exactly how others experience themselves and the world around them.

Matthieu: We may not be able to bridge the gaps between the way things appear and the way things are in all cases, such as in an optical illusion. Other kinds of gaps, however, perhaps more essential ones, can be bridged. For Buddhism, bridging the gap has essentially a pragmatic goal: to become free from suffering because suffering necessarily arises when one's perception of the world is deluded and at odds with reality.

IS COGNITIVE DELUSION INESCAPABLE?

Wolf: Let's pursue further the possibility of distinguishing between appearance and reality. You seem to assume that this epistemic gap can be bridged under certain conditions, whereas I would tend to deny this, maintaining that perceiving is always interpreting and hence attributing properties to sensory signals. In this sense, perceptions are always mental constructs.

Matthieu: Sure. I fully agree with you about sense perceptions. But it is not the same when one engages in an investigation of the ultimate nature of reality. This is why it is important to clarify the domains of knowledge in which it is or is not possible to bridge the gap between the way things appear and the way they are. When we think, "This is truly beautiful" or "This is intrinsically desirable or detestable," we are not aware that we project these concepts onto outer phenomena and then believe that they intrinsically belong to them. This gives rise to all kinds of mental reactions and emotions that are not attuned to reality and will therefore result in frustration.

Imagine a fresh rose that has just bloomed. A poet finds it exquisitely beautiful. Now imagine that you are a small insect nibbling on one of its petals. It tastes so good! But if you are a tiger standing before this rose, you are no more interested in the rose than in a bale of hay. Imagine that you are the rose at the atomic level: You are a whorl of particles passing through nearly empty space. A quantum physicist will tell you that these particles are not "things" but "waves of probability," arising in the quantum void. What is left of the rose as a rose?

Buddhism calls phenomena *events*. The literal meaning of *samskara*, the Sanskrit word for "things" or "aggregates," is "event" or "action." In quantum mechanics, too, the notion of object is subordinate to a measurement, hence to an event. To believe the objects of our perception are endowed with intrinsic properties and autonomous existence is, to take again a comparison with quantum physics, like attributing local properties to particles that are entangled and belong to a global reality.

Buddhist thinkers believe that by using a proper method of investigation, one can fathom, intellectually and experientially, the correct nature of phenomena and free oneself from a mistaken, reified, and dualistic apprehension of reality. Recognizing clearly the mechanisms through which we delude ourselves and adopting a view that is much more attuned to the true nature of phenomena is a liberating process based on wisdom. It doesn't mean that one is not going to be fooled by optical illusions anymore, but that one will not be fooled into thinking that phenomena exist as autonomous, permanent entities.

Wolf: This is interesting because when you talk about delusions, you add qualities to them, emotional qualities such as disgusting or beautiful, attractive or repulsive. I had in mind perceptual delusions for which we can obtain objective data by using a physical measuring device in addition to our perception. Without an independent measure, there would be no way to disclose a delusion as such. This is the way science tries to go about recognizing these delusions and finding out why the brain makes these false interpretations. In most cases where delusions or illusions have been investigated thoroughly, it turns out that they are the consequence of an interpretation or inference that serves the perception of the invariant properties of objects.

Without these mechanisms, for instance, we would not be able to perceive the color of a flower as constant irrespective of the conditions of illumination. The spectrum of the illuminating sunlight changes constantly, and so does the spectrum of the light reflected from an object. Without interpretations, the perceived color of a particular rose would not be the same at dawn and dusk. Our brains correct for this problem. They infer the spectral composition of the illuminating light from the *relations* between the reflected spectra and prior knowledge about the likely color of an object and then compute on the basis of this analysis the actually perceived color of the object. Thus, depending on context, physically different signals may be perceived as similar, and, conversely, physically identical signals may give rise to different percepts. These inferential mechanisms can give rise to "illusions," but they have an exquisitely important function for survival—the extraction of constant properties from an ever-changing world. An animal that uses color to distinguish edible berries from other slightly more violet and poisonous berries cannot rely on an analysis of the "true" or actual spectral composition of reflected light. It first has to assess the spectral composition of the light source—the sunlight—and then must reconstruct the perceived color.

We are completely unaware of the complexity of the computations that ensure constancy and thus survival in our changing world. In essence, all these operations are based on the evaluation of relations. We rarely perceive absolute values such as those that are measured by physical measurement devices, be they intensities of stimuli, wavelengths of sound, light waves, or chemical concentrations. We mostly perceive these variables in relation to others, as relative differences, relative increments, and relative contrasts, and these comparisons are made across both space and time. This is an economical and efficient strategy because it emphasizes differences, permits for coverage of wide ranges of intensities, and, as mentioned, allows for constancy. Given the advantages of these well-adapted mechanisms, it is questionable whether one should call the resulting perceptions "illusions."

Matthieu: I am not speaking of the perceived qualities of objects but of the capacity to dispel cognitive delusion, such as the belief that ugliness is an intrinsic quality of the object you behold. As you said, some illusions help us to adapt to the world, but this is not the case with believing in the existence of permanent phenomena or of an autonomous, unitary self within each person.

Wolf: Now, what is an illusion or a delusion in your view or in a Buddhist view? You stated that even in cases where there is no objective measurement device and one can rely only on one's introspection and perception, it is still possible to distinguish between delusion and reality. I don't see how this is possible.

Matthieu: The perceptual illusions that you describe can be useful to function in the world, as you rightly pointed out, but they don't have much impact on our subjective experience of happiness or suffering. The cognitive delusions I described have an opposite result: They cause us to act in dysfunctional ways that produce suffering. The illusions you are talking about are highly adaptive and can be considered wonders of nature. The ones I mentioned are curses that keep us in a state of deep dissatisfaction. The mind might not be a reliable measuring device or a faithful perceptual device, but it is a powerful *analytical* device. This is true of Einstein's thought experiments, for instance, and it is also true of Buddhism's in-depth investigation of the interdependent, impermanent nature of phenomena.

Wolf: But can you really extrapolate this view to all cognitive functions, to perceptions of social realities, social relations, and belief and value systems? I think the psychophysical examples I just mentioned teach us that we construct what we perceive and tend to experience the result as real. We probably do this not only in the case of visual or auditory perception of tangible objects but also when perceiving social realities.

How can we distinguish between right and wrong when different people perceive the same condition differently, each taking the perceived as real, as correct?

Matthieu: That is why we need logical reasoning and wisdom. If I recognize that no one wants to suffer, then it seems quite straightforward to conclude that harming others is wrong.

Wolf: Sure, but if our intuitions and perceptions of the outer world depend on neuronal processes—and I think there is no way around assuming that the "inner eye," too, is a function of neuronal interactions—then the contents of cognition are determined by the way our brains work and ultimately by genes and postnatal experience. All humans have a rather similar genetic makeup, and therefore we have consensus on many of our perceptions and interpretations. But still we may have rather different experiences, especially when raised in different cultures. Two people observing the same social situation may perceive it in completely different ways. They may come to grossly diverging ethical or moral judgments, unable to convince the other through argument that he or she is wrong because both experience what they experience as the only reality.

The problem is that in the case of the perception of social realities, there are no "objective" measurement devices. There are only different perceptions; there is no right or wrong. This has far-reaching consequences for our concepts of tolerance. Solving such problems with majority votes is clearly no fair solution. Assuming that one's own position is correct and granting others the right to stay with their "wrong" perceptions as long as they do not disturb us is humiliating and disrespectful. Still, this is what is considered "tolerant" behavior. What we should do instead is grant everybody that her or his perceptions are correct and assume that this attitude will be reciprocated. Only if this agreement on reciprocity is violated have the dissenting parties the right to exert sanctions.

I take from what you said that you have a recipe for addressing such cases, and this would of course be of utmost importance to settle cultural conflicts arising from diverging perceptions. Do you think that the mental techniques designed by Buddhist philosophy are able to help in such cases of cognitive dissent, of conflicts between different perceptions that both parties experience as real and true? Can people, through mental practice, find the "correct" solution—in case it does exist—or at least become aware of the fact that the world can be perceived in different ways?

Matthieu: As you rightly pointed out, one must be fully aware of people's ingrained beliefs and moral values and take them into consideration. That being said, social and cultural perceptions can be as deceptive as cognitive delusions, and they are built up in similar ways. We sometimes perceive people from other races, religions, social ranks, and so on as "superior" or "inferior." You might perceive someone as a friend one day and as a foe the next. A person from the Himalayas will probably find most modern art meaningless. All these are mental fabrications. That's where many of our human-made problems arise.

The purpose of the Buddhist approach is not to confront people's views head on by imposing another view that one considers to be superior but to help people see that all such views can be misleading and that we should not casually take them for granted. For example, when refuting belief in the existence of a self, it does not help to merely proclaim, "There is no self." Instead, after having thoroughly investigated the purported characteristics of the self and concluded that it does not exist as a separate entity, one would simply invite others to conduct such an investigation and find out for themselves.

So the idea is not to coerce people into seeing things as we see them or adopt our own aesthetic and moral values and judgments, but to help them reach a correct view of the ultimate nature of things as being devoid of intrinsic reality.

In truth, people from different cultures are all superimposing their particular mental fabrications on reality. The problem can be solved if these people investigate reality through logical reasoning and realize that they are simply distorting reality and that neither the object they are looking at nor the subject who perceives it exists as an independent, truly existing entity. As for "right" and "wrong" and ethical judgments, various forms of conditioning and delusions, as we would say in Buddhist terms, lead to various ethical views and systems—some people think, for instance, that taking revenge on someone, even to the point of killing that person, is ethical. But is it logical to kill someone to show that killing is wrong? Sound reasoning can help to identify some universal principles based on benevolence and compassion that may help us reach a consensus about fundamental values that includes care, openness of mind, honesty, and so forth.

Let's remember that the goal of Buddhism is to put an end to the root causes of suffering. Buddhism considers different levels of suffering in depth. Some forms of suffering are obvious to all: a toothache or, more tragically, a massacre. But suffering is also embedded in change and impermanence: People go for a joyful picnic and suddenly a child is bitten by a snake; someone eats a delicious meal but ends with food poisoning. Many pleasurable experiences soon turn neutral or aversive.

A much deeper level of suffering also exists that we don't usually identify as such and yet is the root cause of all sufferings: As long as the mind is under the influence of delusion and of any afflictive mental state such as hatred, craving, or jealousy, suffering is always ready to manifest itself at any time.

To take the example of impermanence, at each moment everything changes, from the change of seasons and of youth to old age, to the subtlest aspects of impermanence that take place in the shortest conceivable period of time. Once we have recognized that the universe is made not of solid, distinct entities, but of a dynamic flow of interactions among countless fleeting phenomena, it has major consequences in weakening our grasping onto the reality we see before us. A proper understanding of impermanence helps us to close some of the gap between appearances and reality.

EACH PERSON TO HIS OR HER OWN REALITY

Wolf: What is reality? Is it not that people look at the same thing in different ways?

Matthieu: How can one be sure of that? There is no way to prove that a reality exists out there behind the screen of appearances, a reality that exists in and of itself, independent of us and the rest of the world. To assume a substrate beneath appearance may seem rational, but it surely needs to be questioned. Even before the advent of quantum physics, the mathematician Henri Poincaré said, "It is impossible that there is a reality totally independent of the mind that conceives it, sees it or senses it. Even if it did exist, such a world would be utterly inaccessible to us."[7] Simply measuring what we can apprehend in the world does not prove that what we observe exists from its own side and has intrinsic characteristics. Perceptions, appearances, and measurements are just events. While looking at the moon, if one presses one's eyeballs with one's fingers, one will see two moons. One may do that a thousand times, without the second moon being any more real for all our efforts. But we should also ask ourselves whether the first moon is ultimately real in the way it seems to us to be.

This is especially true of the qualities that we ascribe to phenomena. If something could be intrinsically beautiful, independently of the observer—an object of art, for instance—it would strike everyone as beautiful, whether it be a sophisticated New Yorker or a reclusive forest dweller who has never been in touch with the modern world.

Wolf: So you have a constructivist approach, and you consider that everybody constructs the world in his own way.

Matthieu: Yes, and that way is deluded because we all keep on assigning an element of truth to our superimpositions on the world. What Buddhism does is *deconstruct* ordinary perceptions by conducting an in-depth investigation of the nature of what people see to make them understand that they are all distorting reality in different ways.

Wolf: I would not say "distorting" because if there is no objectivity, you can't distort anything—there is nothing objective to distort. People simply give different interpretations.

IS THERE AN OBJECTIVE REALITY "OUT THERE"?

Matthieu: Objectivity is not just one of the many versions of what various people perceive but the irrefutable understanding that *all* phenomena are impermanent and devoid of intrinsic characteristics. This applies to all appearances, all perceptions, all phenomena. Distortion, therefore, is not defined in comparison with a true, self-existing reality. Distortion is to attribute any kind of intrinsic reality, permanence, or autonomy to phenomena.

A nondistorted view is not one of the many ways that things appear to us ordinarily. Rather it is an understanding of the process of delusion, the realization that the phenomenal world is a dynamic, interdependent flow of events and the knowledge that what we perceive is the result of the interactions of our consciousness with these phenomena. That understanding is correct in all situations.

Buddhist texts use the example of a glass of water. They point out that water could be perceived in countless different ways. We perceive it as a drink or something to wash with, whereas for a fish water is like space. Water is known to terrify someone stricken with rabies, whereas it appears to a scientist like a great number of molecules. According to Buddhist cosmology, some sentient beings may perceive water as fire and some as delightful ambrosia. However, behind all these perceptions, is there a true, self-defined glass of water? The Buddhist answer is no.

When a complex set of phenomena interacts with our senses and consciousness, a particular object crystallizes in our mind. We might see this object as something to drink or we might find it utterly terrifying if we are suffering from rabies. At no point in time and space can one find autonomous objects or subjects existing in and of themselves. The glass of water has never been there on its own, endowed with a true, separate reality. It only exists in a world of relations. What Buddhism calls "reality" is not phenomena such as self-existent water but the realization of their impermanence and lack of intrinsic reality.

Wolf: This notion fits well with the constructivist position of contemporary neurobiology. Still, the world has certain properties, and animals seem to share the same criteria for the definition of objects and qualities. In our ordinary mesoscopic world, solid, nontransparent objects are called rocks behind which animals can hide, roll down if placed on a slope, and so on. Evidence suggests that all mammals use similar *gestalt* principles (i.e., similar rules and hypotheses to construct their percepts).

Matthieu: Sure. Different types of consciousnesses that are similar enough—those of human beings and some animals (great apes, dolphins, elephants, and others)—will perceive the world with a corresponding degree of similarity. The more the structures of these consciousnesses differ, the more their world is different. The point is that when you free yourself from cognitive delusions first through an analytic investigation and then through integrating the resulting understanding into your way of relating to the world, you will gradually gain freedom from the compulsive attraction and repulsion that usually result from delusion. Thus, as you get closer to understanding the true nature of phenomena, you get closer to understanding the root causes of suffering and to freeing yourself from these causes. This freedom brings about a more optimal way of being that is much less susceptible to suffering.

Wolf: This is interesting: First you take a constructivist stance and then you question the validity of your constructions and conclude that this epistemic turn, this switch in your cognitive approach, leads to the reduction of suffering.

Matthieu: That's the goal of the Buddhist path.

Wolf: Let me try to understand this. I think many of us who were raised after the Age of Enlightenment assume that suffering can be reduced if we find out how things work and how we can manipulate them for improving our condition. For this strategy, it is imperative to be able to distinguish among delusion, false beliefs, and superstition, on the one hand, and valid interpretations, on the other.

In medicine, for example, we try to find causal relations between events, identify infectious agents, and then develop treatment. These concepts are competing, however. Adherents of allopathic medicine would agree that you need a certain dosage of an antibiotic to make it work. Homeopaths, in contrast, would maintain that it is dilution that matters more than the drug, even if the dilution is so high that it becomes highly unlikely that a single molecule of the drug is actually left in the bottle you buy at the pharmacy. They claim that the treatment is still effective because the water or the pill is supposed to keep the memory of the molecules that had been there before dilution.

We could leave it there rather than trying to find out which treatment is more effective. Or we could perform a double-blind study[8] and discover that placebos are as effective as the homeopathic treatment whereas the antibiotic is much more efficient. In doing so, we assign a property to drugs and verify by experiment that the property is causally related to an effect and therefore with all likelihood relevant to the effectiveness of the treatment. By repeating the experiment and establishing dose-response curves, we establish causal relations and ensure that the property of the drug is constant. However, if I understand your approach correctly, you would deny that the drug has this invariable property, or you would say that it has the defined property only in the special context of this assay.

Matthieu: That is not exactly what I meant. What you describe refers to the difference between what Buddhism calls *correct relative truth* and *erroneous relative truth*.

Wolf: Explain that to me.

CAUSALITY AS A CORRELATE OF INTERDEPENDENCE

Matthieu: In Buddhism, *absolute truth* refers to the recognition that phenomena are ultimately devoid of intrinsic existence. *Relative truth* is to acknowledge that these phenomena arise not in haphazard ways but according to the laws of causality. Far from refuting them, Buddhism is based on these laws. It even emphasizes that these laws are ineluctable and should be understood and observed if one wants to escape suffering. Of course phenomena do have relative properties that allow them to act on other phenomena and be acted on by them through mutual causation. However, according to Buddhism, to believe that penicillin is *intrinsically* good, no matter what the circumstances might be, is incorrect. For instance, some people happen to be allergic to penicillin. Although in most situations penicillin is beneficial for people at a correct dosage, it's not beneficial *in itself* because it can be poison for someone who is allergic to it.

What Buddhism concluded after investigating the fundamental nature of the phenomenal world is that these properties are not intrinsic to the object but arise through particular relations between phenomena. Heat can only be defined in relation to cold, high in relation to low, the whole in relation to its parts, a mental concept in relation to its base of designation, and so on. The same substance could be curative to someone and poisonous to someone else or, like digitalin for the heart, curative in small quantities and poisonous in large quantities. This is also true at the fundamental level of quantum physics, with the absence of local properties in elementary particles, a notion that bothered Einstein and yet has since been proven to be true.

This is all the more true when it comes to judgments, such as beautiful and ugly, desirable and hateful. Yet ordinarily we do reify the world, attribute intrinsic properties to what is around us, and react accordingly in dysfunctional ways that ultimately cause suffering. That is what I meant. Valid cognition should withstand the most thorough and in-depth analysis. Apparent properties of phenomena don't. There is a difference between apparent, relative, conditioned properties and intrinsic ones, but typically we ignore it. This is not a mere intellectual distinction—ignoring it causes us to act in ways that stand at odds with reality and are, therefore, dysfunctional.

Wolf: When you conduct "thorough and in-depth analysis," does that include experiments?

Matthieu: It can, of course. In the example that you quoted, the experiment would consist of conducting a double-blind study looking at the effect of penicillin on a large sample of individuals and doing the same with homeopathic medicine. This experiment would lead to the conclusion that you mentioned, which is that penicillin is an active substance, whereas homeopathic remedies are not less but also not more efficient than placebos. That would be considered to be correct relative truth, that is, valid knowledge of the phenomenal world. But no matter what the results of the study are, *ultimately* penicillin is still an impermanent phenomenon, the properties of which differ depending on circumstances.

Wolf: Is it then that we should detach ourselves from the reality that we perceive, including the social reality with its ethical and moral value systems? Should we abandon the belief that there is anything reliable and constant out there and content ourselves with the insight that "in reality" there are no invariant properties attached to anything, neither to inanimate objects nor to plants, animals, or persons? What do we gain if we abandon the idea that things have invariant properties that allow us to recognize and categorize them and instead adopt the view that properties are merely assigned and permanently changing in a context-dependent way? To avoid the attribution of properties is one strategy to avoid falsification because right and wrong lose their antinomy. However, I cannot see why this relativism should reduce misconceptions and suffering.

Matthieu: What you describe would rather fall into the extreme of nihilism. Buddhism clearly acknowledges the workings of the laws of causality and accepts the idea that, at the relative level, some properties or characteristics of some phenomena might endure for a while so that we may rely on these characteristics to function in our daily life in a coherent way. Stone and wood remain solid long enough, for instance, for us to build our houses with them. But it still remains that they are fundamentally impermanent. The same piece of wood that makes a useful chair for us is desirable food for a termite. Wood will eventually disintegrate into dust. Even now, wood is not fundamentally made of "wood" stuff but of particles or quarks; to view it from the perspective of quantum physics, it is the result of ungraspable quantum events. It all depends on how you look at it. Although we can use a tool made of wood for a long time in a skillful and useful way, ultimately it is devoid of intrinsic existence. So the relative, conventional truth is not in opposition with the ultimate truth about the nature of phenomena. The latter is simply the ultimate nature of the former.

To conclude that phenomena are impermanent and interdependent is the only outcome of a careful, logical investigation. Any other proposition stating that there must be entities endowed with permanence and intrinsic properties, whether that might refer to atomic particles, the concept of beauty, or the existence of a creator god, cannot withstand such a thorough investigation.

Why would it reduce suffering? When you take the way things appear as being reality, what Buddhist scriptures call "happily taking things for granted without any analysis and investigation," you are heading for trouble because being at odds with reality will inevitably lead to some kind of dysfunction. For instance, when you cling strongly to something, assuming that it will last, that it is truly yours, and that it is in and of itself desirable, you are not only at odds with reality but putting yourself in a vulnerable situation because all your relations to the object of your grasping are warped. Frustration and suffering will ensue when it turns out that the object is in fact impermanent, it can be destroyed or lost, and it can never be truly yours. It may also suddenly appear to be undesirable simply because your projections onto it have taken a U-turn.

However, if you think, "Phenomena appear as interdependent events devoid of autonomous, inherent characteristics and existence," because such understanding is congruent with reality, you are much less likely to relate to objects in ways that lead to disappointment and suffering.

CONSTRUCTING AND DECONSTRUCTING REALITY

Wolf: Your epistemic scheme resembles the position of radical constructivism. Brains construct their views of the world on the basis of inherited and acquired knowledge. Because different brains have different knowledge bases, they may arrive at different views. Neurobiologists would agree so far. We perceive the world as we do because our brains are the way they are. Because the genetically and culturally transmitted cognitive schemata (priors) are quite similar, we tend to perceive the world in similar ways. You go one step further, however: You stated that one major cause of suffering is that people are not aware of these facts and continue to believe that perception reflects reality. Consequently, if perceptions differ, the respective other is perceived as being wrong and attempts to correct the apparently wrong perception inevitably cause suffering on both sides. Is the Buddhist position that neither right nor wrong exists because the conflict arises only between conceptual attributions that should not have been made to begin with and that there is therefore no point in trying to convince each other through argument?

Matthieu: Not quite. When people holding various opinions and entertaining various perceptions deconstruct their respective delusions, they cannot but agree on the correct understanding of the ultimate nature of phenomena.

Wolf: Okay, but they would not be able to deconstruct their perceptions. For them, what they perceive is real. They would, however, agree at a metalevel and consent that, irrespective of our idiosyncratic perceptions, the objects of the perceivable world are impermanent, devoid of intrinsic qualities, and only defined in terms of relations.

Matthieu: People may still perceive as reality something that is, for instance, nothing more than an optical illusion, but at the same time they will recognize that this illusion does not reflect the true nature of the object perceived. The goal is not to agree on sensory perceptions but to understand that these perceptions result from constructing a fictitious reality. All parties can free themselves from cognitively deluded ways of apprehending reality.

Wolf: In other words, they would continue to see what they see, but they would become aware that this is not the only way it can be seen. This epistemic stance also pervades most of the occidental philosophical schools and agrees perfectly well with what we know about the neurobiological underpinnings of perception.

Matthieu: Yes, that's right, but it does not stop there. They would further acknowledge that their way of seeing is fabricated. Analytical meditation and mental training would allow them to recognize that their habitual tendencies cause them to attach various qualities to objects even though these qualities are not invariable attributes of the objects. Thus, through training, insight deepens, and one can come to understand the constructed nature of the cognitive processes that take place in our minds. This, in turn, makes it easier to detach oneself from grasping, attraction, and repulsion and achieve greater inner freedom.

Wolf: I find the idea fascinating that the brain, by cultivating insight, should be able to arrive at a level of metacognition that allows it to discover the nature of its own cognitive processes. By "metacognition" I mean a process by which the brain applies its cognitive abilities not to the investigation of objects in the outer world but to the investigation of its own operations. The architecture of the human brain could well allow for such metacognition because its multilayered organization permits it, in principle, to iteratively subject its own processes to scrutiny.

Matthieu: You can perceive something as being permanent while understanding at the same time that it is utterly transitory. In this way, you can cease to attribute solid, unchanging qualities to what you see. Consequently, you would not react in deluded ways. To deconstruct the world of appearances has a liberating quality. You are no longer entangled in your perceptions and cease to reify the phenomenal world. This has profound impacts on the way you apprehend the world and consequently on your experience of happiness and suffering.

When all mental fabrications are unmasked, you perceive the world as a dynamic flow of events, and you stop freezing reality in various deluded ways. Take the example of water and ice. When water freezes, it forms solid shapes that can cut your hand or break your bones if you fall on it. Now, you could think that this is the nature of water: It has a particular shape, it is hard, and so on. You could also make various forms out of ice—flower, castle, statue of a loved one, or representation of a deity. I have even heard music played on instruments made of ice! But with just a little heat, all these different, well-defined forms melt into the same fluid, shapeless water. If you remain mindful of this at all times, then when you see a flower made of ice, you are fully aware of its impermanent nature, that there is nothing intrinsic in either its "flower" quality or its "beauty." Water is neither a flower nor a castle nor a god. It is a dynamic flow that can momentarily assume seemingly stable configurations. Likewise, if we don't freeze reality, we will not be caught in reifying it as something solid, endowed with true, intrinsic existence, and we will not be deluded.

Wolf: Water is indeed a nice metaphor: A river is never the same at two different moments. This represents the experience of an ever-changing world that will never ever come back to the point where it was before. The same holds true for the brain. It is also constantly changing and will never come back to the same state. This permanent and never-repeating flow of changing states is probably the reason that we perceive time as directed.

But why do you say "true"? Why should the ever-changing water be truer than a statue of the Buddha made out of ice or stone? The distinction among ice, liquid water, and vapor is fundamental to the understanding of the properties of matter as it describes different aggregate states of the same molecular constituents. What is the meaning of "true" here?

Matthieu: One is not truer than the other. Neither water nor ice is endowed with true existence. What is true, however, is that all these aspects—solidity, fluidity, form—are impermanent. You could deconstruct water into molecules and particles and particles into merely interconnected events, quantum probabilities, not "things" standing on their own.

Such understanding comes from deeply investigating the way our own mind functions and requires a mind that is clear and stable enough to follow a rigorous process of introspection and make proper use of logic to deconstruct our naïve perceptions of reality.

Wolf: I consider it quite surprising that contemplative techniques lead to insights into the nature of the world that actually contradict our primary perceptions.

Matthieu: Even in the domain of physics, Einstein's thought experiments and his visionary insights led him to formulate the theory of relativity, which also contradicts our primary perceptions. The same is true, even more so, with quantum physics.

Wolf: Intuition and introspection have not proven to be particularly effective tools when it comes to understanding phenomena not directly accessible to our senses. This even holds for the organization and functions of our own brain, an epistemic problem that we shall have to discuss at some stage. In science, we have certain rules or strategies to validate what we believe to be the case. We postulate reproducibility, predictability, consistency, and an absence of contradictions. Sometimes we even apply aesthetic criteria such as beauty because we consider simple explanations as more trustworthy or powerful than complex ones. I can see the point that one can obtain insights through the investigation of the way one's own mind functions. But how can one validate this process? How can one communicate to others how reliable one's own introspective evidence is? What is "right" in this context of self-inquiry? I would like to learn more about this.

REFINING THE TOOLS OF INTROSPECTION

Matthieu: Let's take the example of the telescope. There are two reasons that you would not be able to see something clearly through a telescope: either the lens is dirty or not properly focused or the telescope is unstable and shaking. So when you lack either clarity or stability (or both), you can't see the object properly.

Wolf: In the case of the telescope, it's clear. There are objective criteria for what constitutes a sharp, high-contrast image. Modern cameras use these criteria to adjust the focus automatically. But what are the criteria in the case of introspection? How does it feel when the cognitive system is properly adjusted?

Matthieu: Well, similarly, you need to make the mind's telescope more focused,[9] clear, and stable. Introspection has long been discredited because the subjects who were asked to engage in it in laboratory studies did so with minds that were distracted most of the time. Distraction creates an unsteady mind. In addition, an untrained mind lacks the limpid clarity that allows one to see vividly what is happening within oneself. So whether the mind is carried away by distractions or sinks into a cognitive opacity, it will not be able to pursue proper introspection.

Wolf: So stability and clarity would be the two main criteria?

Matthieu: Yes. To use your expression, it certainly "feels" different when the mind is agitated, distracted, and murky versus when it is stable and crystal clear. A clear and stable mind brings not only inner peace but deeper insights into the nature of reality and the mind itself. These effects are not just imaginary but well-defined mental states that can be experienced again and again.

Wolf: Does reproducibility also figure as a criterion, just as in the scientific approach?

Matthieu: Whenever a contemplative whose mind is not constantly carried away into a whirlpool of thoughts investigates a particular aspect of mind through introspection, he usually comes to similar understandings and insights, which is not the case for one whose mind is wild and deluded.

Wolf: In scientific terms, that would be noise reduction—stabilization of your cognitive system.

Matthieu: That's right. You need to get rid of both mental cloudiness and agitation.

Wolf: Is this a quality your "inner eye" can acquire?

Matthieu: Yes, this is the result of sustained practice. Many techniques and meditative practices can help you to progress toward achieving a clear and stable mind. When you thus train your mind, you also perceive and understand mental phenomena more accurately. At the same time, you realize that the boundary between mental events and outer phenomena is not as solid as it seems. This approach is phenomenological because what you are investigating directly is your experience. What else could you investigate anyway? Your experience is your world.

Wolf: Is your advice to consolidate your insight and resist contextual influences?

Matthieu: You will of course still perceive all outer and inner phenomena. You will actually perceive them more vividly and with a penetrating insight because you will have stopped superimposing mental fabrications onto them. This allows for an interindividual consensus among trained contemplatives, which is not the case for untrained subjects. We may compare this process to that followed by mathematicians who, having undergone common training, can understand each other as they reach similar insights and speak the same language. Unless you are trained in mathematics, it would be hard to grasp their concepts and follow their discussions. Likewise, trained contemplatives come to similar conclusions about the nature of mind and about the fact that phenomena are impermanent and interdependent. This verifiable intersubjective agreement confers validity on their understanding.

FIRST-, SECOND-, AND THIRD-PERSON EXPERIENCE

Wolf: Mathematicians can take a pencil and write down a formula and then develop a proof using logical rules. How can you, with your teacher—I assume you need a teacher to tune your instrument, your inner eye, your microscope—know that you are going the right way?

Matthieu: The way to know that is through what Francisco Varela, Claire Petitmengin, and others in the field of cognitive science call the second-person perspective, in complement with the first- and third-person perspectives. The second-person perspective involves an in-depth, properly structured dialogue between the subject and an expert who leads the dialogue, asking appropriate questions and allowing the subject to describe his or her experience in all its minute details.

In the Tibetan tradition, for instance, the meditator will from time to time report about his meditative experiences to his teacher. The difference here is that the second person is not just a skillful psychologist but someone who has a deep experience of meditation achieved over many years of dedicated practice and whose experience has culminated in profound, clear, and stable insight on the nature of mind, something often referred to as spiritual realization or accomplishment. On the basis of his genuine and vast experience, a qualified teacher will be able to appraise the quality of the student's meditative practice and see whether it reflects genuine progress or mere self-deceptions.

You could argue, "From a third-person perspective, how can I verify the validity of such judgments?" Well, you can actually verify all this by yourself, but not without training. A similar process exists in science. If you don't know much about physics and mathematics, then you begin by trusting the experts because you assume that they are reliable. Why should you believe them? In the beginning, you trust them because they agree among themselves after having carefully verified each other's findings. But you don't have to stop there. You also know that if you were to train properly in their discipline, then you would be able to check all this by yourself. You don't have to trust these experts forever on the basis of blind faith, which would remain unsatisfactory.

Meditation is not mathematics but rather a science of the mind, and it is conducted with rigor, perseverance, and discipline.[10] Thus, when experienced contemplatives come to similar conclusions about the workings of the mind, their cumulative experience has a weight comparable to that of expert mathematicians. As long as you have not personally involved yourself in your own investigation and experience, a gap will always exist between what you are told and what you know through direct experience, but that gap can be gradually bridged by enhancing your own expertise.

Wolf: Doesn't this apply to all strategies of gaining insight? You have to agree on certain criteria, you have to agree on certain procedures for the acquisition of knowledge, then you explore, and finally you sit together and try to find out whether you reach a consensus.

Matthieu: It seems appropriate to speak of a "contemplative science" because these are not vague descriptions based on mere impressions. In the Tibetan contemplative literature, one finds entire treatises that describe various steps for analyzing the mind and offer a detailed taxonomy of the various kinds of mental events. They also describe thought processes, how concepts are formed, what the qualities of pure awareness are, and so on. These treatises also teach meditators how to avoid misconstruing fleeting experiences as genuine realization. All these phenomena are described by people who have acquired a penetrating insight into what is going on in their minds. You could argue that they are all deceiving themselves, but it would be a bit strange if a whole cohort of people with sharp acumen who have refined their introspective faculties to such an extent deluded themselves in exactly the same way at various times in history and in various places, whereas untrained people with wild, confused minds somehow had a more reliable picture of the workings of the mind.

That's why the Buddha encouraged contemplatives to practice assiduously by saying, "I've shown you a path, and it's up to you to travel it by yourself. Don't believe what I say simply out of respect for me, but examine the truth of it very thoroughly, as when examining the purity of a piece of gold by rubbing it on a flat stone, beating it, and melting it." We should take not things for granted without verifying them for ourselves.

Some things are not accessible to your knowledge. Some elude your direct experience forever. When believers of theistic religions speak of the "mystery of God," for instance, they accept the fact that they will never fully know the nature of God through their limited, imperfect experience. In the case of Buddhism, it is said that other aspects of reality are not accessible to your present cognitive capacities but that are by no means inaccessible forever. What is not clear to you now can become completely clear in the future through investigation and training.

Wolf: This would also imply that those who have not adopted these practices should not rely on their firsthand intuitions, their immediate internal judgments, because they haven't tuned their instrument. They all have to be considered naïve and deluded by blurred perceptions.

Matthieu: Yes, but they have the potential to do so. They remain naïve as long as their potential for understanding remains untapped.

Wolf: Because most of us have untrained minds, because only a small minority has gone through the process of tuning the inner-eye microscope, this situation seems fairly depressing. In the West since the Age of Enlightenment, we have focused on science as the definitive source of knowledge, but because we are all naïve, the great minds—Plato, Socrates, Kant—included, how can we trust our findings and conclusions? This reminds me of a curious conundrum that illustrates how insights derived from introspection and scientific inquiry can diverge. Consider the many and widely differing theories about the organization of our brains that have been derived from introspection and observation of the behavior of the respective other. Most of them turn out to be incompatible with what we observe once we start to examine cognitive functions with quantitative psychophysical methods and subject brains to scientific investigation. To the best of my knowledge, this holds for all prescientific theories of the brain, whether formulated in the framework of Eastern or Western philosophies. Does this discrepancy imply that our intuitions are simply naïve because they are untrained? In this case, one would expect that those who have experience with mental practice, who have tuned their inner eye, might come to more valid conclusions about the way their brains function. Trained Buddhist minds should experience the working of their brains in a more "realistic" way than that suggested by our untrained Western naïve intuition.

Matthieu: It would be more correct to say that trained contemplatives experience the working of their *minds* in a more realistic way than untrained people. This does not mean that you will relate your experience with specific areas of the brain as an fMRI machine would do. Whether you are naïve about the functioning of your mind or a trained meditator, in both cases, as you know well, you normally cannot even feel your brain, much less know what is going on in its various areas and networks. That being said, the collaboration between contemplative and brain scientists that has unfolded over the last 15 years or so has shown clearly that these fields can mutually enhance each other's understanding and correlate first- and third-person perspectives.

Through pursing a first-person approach, a contemplative will not find out directly which areas of the brain are involved in compassion or focused attention. However, a trained contemplative will be highly aware of his cognitive processes, of the way thoughts unfold, and of the way emotions arise and how they can be balanced and controlled. The meditator will also have some experience of what is known as *pure awareness*, which is a clear and lucid state of consciousness devoid of mental constructs and automatic thought processes. The meditator may also understand that there is no such thing in the mind as a central, autonomous self, which I think fits quite well with the views of neuroscience.

The recent development of what is now called contemplative neuroscience explores how this contemplative knowledge and mastery of the mind relates to specific brain activities and how the meditator may or may not be able to monitor and control these at will. The experience of contemplatives can also be harnessed to better interpret findings about the workings of the brain, particularly in the field of emotions, well-being, depression, and other heightened states of mind.

A PHYSICIAN AND A CURE

Matthieu: To say that most of us are somehow deluded is not fundamentally a pessimistic attitude because there is a way out of our delusion; we are not stuck here. The doctor who diagnoses a sickness or an epidemic is not pessimistic. He knows that there is a big problem, but he also knows that this problem has causes that can be identified and a cure for these ailments may exist. In the Buddhist scriptures, one often compares the Buddha to a skillful physician, sentient beings to sick patients, the teachings as the doctor's prescription, and the practice of these teachings as the treatment. The main reason not to be depressed is that the mind has the potential to change and cure itself of delusion, so as to perceive reality as it is.

Wolf: The reality you are talking about is then essentially an internal state devoid of delusions and misconceptions because there is no such thing as a true outer reality.

Matthieu: It is the reality of recognizing the nature of pure awareness, as well as the nature of suffering and its causes—the mental toxins—and the possibility of getting rid of these causes through cultivating wisdom. But it is also apprehending outer reality in a more correct way, as interdependent events devoid of intrinsic existence. It is not about gathering detailed information about everything through exploring the intricacies of natural phenomenal realities, as science does, but rather about understanding the fundamental nature of phenomena to dispel basic ignorance and suffering.

Not all knowledge has the same utility in terms of dispelling suffering. A curious person kept asking the Buddha endless questions on all kinds of subjects, such as whether the universe was finite or infinite. The Buddha often kept silent instead of answering. On one of these occasions, he took a handful of leaves and asked his visitor, "Are there more leaves in my hand or in the forest?" The inquisitive person was a bit startled but nevertheless answered, "Of course there are fewer leaves in your hands." The Buddha then commented, "Likewise, if your goal is to put an end to suffering and reach enlightenment, some kinds of knowledge are useful and necessary, while others are not." Many things might be quite interesting in and of themselves (e.g., knowing the temperature of stars or the way flowers reproduce) but are not directly relevant to freeing yourself from suffering.

Not all information is equally useful. It also depends on your purpose. Valid knowledge about the process of cognitive delusion is immensely useful if one falls prey to compulsive attachment or hatred because this will help dispel suffering.

THE ETHICS OF PRACTICE AND SCIENCE

Wolf: I think we are in agreement that one cannot derive ethical values from scientific exploration alone. Science helps us to distinguish between correct and incorrect interpretations of the observable, but it does not liberate us from the burden of making ethical judgments.

Matthieu: This is understandable because those ethical values are not the primary goal of science. Knowledge obtained through scientific inquiry has no moral value on its own. It is the way we make use of such knowledge that morality comes in. By contrast, the primary goal of Buddhism is to get rid of suffering, which is obviously linked with ethics.

Wolf: Is it possible then to derive values from introspection, from mental practice? To me it appears that values emerge from collective experiences and become condensed in either religious commandments or systems of law. Communities found out, by trial and error over generations of experimentation, which attitudes either reduced or increased suffering. They then extracted rules of conduct and codified their cumulative experience. These rules were either projected as God's will to increase their authority and the community's compliance or were incorporated in legal systems. In both cases, remuneration and punishment are common tools to obtain obedience.

Matthieu: It can be true in some cases. In Buddhism, which invokes no divine authority, ethics is a set of guidelines derived from empirical experience and wisdom to avoid inflicting suffering on others and yourself. The Buddha is not a prophet, a God, or a saint but rather an awakened one. Ethics is really a science of happiness and suffering, not a set of rules proclaimed by a divine entity or dogmatic thinkers. Because ethics is all about avoiding inflicting suffering on others, having more wisdom and compassion, together with gaining a better understanding of the mechanisms of happiness and suffering and the laws of cause and effect, will foster ethical systems and practices that are more likely to fulfill their purpose.

THREE ASPECTS OF BUDDHIST PHILOSOPHY

Wolf: So far we have been dealing with three aspects of Buddhist philosophy; please correct me if I have misunderstood something. One is the philosophical, epistemic position of Buddhism, which is clearly a rather radical, constructivist position that declares most of what we perceive outside of our own mind, and for untrained, naïve humans, also most of what one experiences with one's inner eye, as delusive.

The second aspect is the conviction that it is possible to fine-tune one's inner eye through practice to experience what one's mind and reality are all about.

Finally—and this seems to be the most important point and the consequence of the first two—if the goal to purify one's mind is achieved and perception is no longer contaminated by false beliefs, then one changes basic traits of one's personality and thus becomes a better person who can contribute more effectively to the reduction of suffering.

Thus, Buddhist philosophy is partly a sophisticated science of cognition and partly a pragmatic educational system. Unlike Western epistemology, however, it is considered to be an experimental discipline that tries through practice and mental training to clarify the conditions of our cognition and thereby to discover the essence of reality: first fine-tune your internal microscope and then learn about the world.

Western science disclaims that it is able to derive any moral values or formulate rules of conduct based on its observations. All it claims is that ethical choices usually yield better results if they are based on secure evidence and guided by rational arguments rather than by beliefs, superstitions, or ideological dogmas. It further promises that suffering can be reduced by identifying its causes and developing tools for their eradication.

Buddhist philosophy also claims to apply the criteria of an experimental science, but it goes one step further by promising that one is able to derive values from its practice, that practitioners are transformed for the better, and that suffering is reduced in this way.

Matthieu: Well said. Yes, an ethical dimension is embedded in the whole Buddhist approach because knowledge is used to relieve suffering. For this one needs to distinguish the kinds of actions, words, and thoughts that will cause suffering from those that will bring fulfillment and flourishing. There is no such thing as absolute good and evil, only the suffering and happiness that our thoughts, words, and actions bring about for others and ourselves.

Values can also be related to a correct understanding of reality. We would say, for instance, that the pursuit of selfish happiness is not attuned to reality because it assumes that we can function as discrete, separate entities minding our own business, which is not the case. We are bound to fail in that pursuit. On the contrary, understanding the interdependence of all beings and phenomena is the logical ground for growing altruism and compassion.[11] To endeavor in achieving your happiness and that of others simultaneously is more likely to succeed because it is attuned to reality. Selfless love reflects some understanding of the intimate interdependence of all beings and all beings' happiness, whereas selfishness exacerbates individualism and increasingly widens the gap between others and us.

Wolf: You say that these negative attributes are actually deluded perceptions of reality. So you argue in a sense, like Rousseau, that reality is good in its essence.

Matthieu: Reality is neither good nor bad, but valid and invalid ways of apprehending reality exist. These various ways have consequences: A mind that does not distort reality will naturally experience inner freedom and compassion, instead of craving and hatred. So, yes, if you are attuned to the way things really are, then you will naturally adopt behavior that will be conducive to less suffering. Mental confusion is not only a veil that clouds our understanding of the true nature of things. Practically speaking, it also prevents us from identifying the kind of behavior that would allow us to find happiness and avoid suffering.

Wolf: Obviously, if you act in agreement with the true conditions of the world, then you will have less trouble because there will be less contradiction, less conflict—

Matthieu: Right. That is why the investigation of the nature of reality is not just pure intellectual curiosity: It has profound repercussions on our experience.

Wolf: Hence, one should sharpen one's cognitive tools to perceive reality more correctly.

Matthieu: If you recognize that reality is interdependent and impermanent, then you will adopt the right attitude and be much more likely to flourish. Otherwise, as Rabindranath Tagore wrote, "We read the world wrong and say that it deceives us."[12]

Wolf: We could also express this in Darwinian terms: If one's model of the world is correct, then one will experience fewer contradictions, make fewer false judgments, cope better with the fallacies of life, and inflict less suffering. Thus, one should try to get a realistic model of the world. I think all cultures have in common the urge to try to understand the world, but the motives and strategies differ. Reducing suffering is certainly one goal, but there are others, too—those who know have better control over the world, they can dominate others, they have privileged access to resources. Realistic models of the world increase fitness.

One way to obtain knowledge is science. The resulting insights can be made explicit and the reasons can be made transparent. Another source of knowledge is collective experience. The insights obtained in this way often remain implicit—one knows, but the origins of one's knowledge remain opaque. Then there is the strategy that you have explained, which attempts to use introspection and mental practice to learn about one's condition. Finally, there is the evolutionary, pragmatic strategy to arrive at better models of the world. As creative creatures, we have the option to imagine and test models. We can then select those that work best for us, adhere to those that reduce suffering, and drop those that increase it.

Matthieu: This is crucial.

Wolf: And you can optimize your models—

Matthieu: —through analytical meditation, logical investigation, and valid cognition. Going back and forth between your inner understanding and being confronted with the outer world, you will be able to integrate this deeper understanding into your way of being.

Wolf: The implication is then that the brain can impose on itself a training procedure that induces lasting changes in its own cognitive structures?

Matthieu: That's why a mere theoretical understanding will not work. Training implies cultivation, repetition that leads to slowly remodeling your way of being, which will be correlated with a remodeling of your brain. You need to acquire a correct understanding and then cultivate that understanding until it becomes fully part of yourself.

Wolf: It's interesting to investigate how this process can be initiated. You probably need a teacher who tells you that there is something to discover, or is there an internal drive, built into our brains, that motivates self-exploration and promotes self-improvement?

Matthieu: The internal drive arises from a deep aspiration to free oneself from suffering. This aspiration, in turn, reflects the potential that we have for change and flourishing. A qualified teacher plays a crucial role in showing and explaining to us the means to achieve that change, in the same way that the guidance of an experienced sailor, craftsman, or musician is invaluable for those who want to learn those skills. You may wish to reinvent the wheel, but it is senseless not to benefit from the vast accumulated knowledge of those who have mastered their arts and skills, such as mountaineers who have climbed the highest peaks and sailors who have navigated the seas for 40 years. Wanting to learn everything from scratch without benefiting from others' wisdom is not a good strategy. Many generations of sailors have tried all kinds of ways of navigating and have drawn maps of the many places they have visited. Likewise, Buddhist contemplative science has 2,500 years of cumulated empirical experience of investigating the mind, beginning with Buddha Shakyamuni, and it would be silly to ignore it.

Wolf: I begin to understand you. Longing for happiness and minimizing suffering is the drive, but to get there, you need to make your internal model concordant with the "real" conditions of the world. Accordingly, we should be able to identify some of these wrong views and wrong ways of thinking.

Matthieu: To eliminate wrong views is one of the chief goals of the Buddhist path.

A SUMMARY

Wolf: Matthieu, we had a wonderful morning. I would just like to recapitulate what we discussed. We had an in-depth discussion of epistemic questions comparing Western and Buddhist sources of knowledge, the latter being mainly introspection, mental practice, and observation of the world after having purified one's mind...

Matthieu: ...and pursued an analytical approach to reality.

Wolf: An analytical approach that requires one to first fine-tune the inner eye of the mind to purify one's own cognitive system. This, as I understand it, has far-reaching consequences. One consequence is that it allows one to avoid taking for granted what one perceives and helps one to perceive reality as something that is transitory, not endowed with fixed, context-independent properties. This in turn would permit construction of more realistic models of reality, reduce conflicts between false models and reality, and thereby lead to a reduction of suffering.

One important aspect of this is to learn to disengage to avoid emotional grasping. By realizing that objects have no fixed qualities per se, we detach those qualities from our understanding of the objects. If I understood you correctly, this also applies to the emotions that we associate with social situations and other sentient beings. Clinging and attachment act like distorting filters on one's perceptions that prevent us from perceiving the real world—and should therefore be avoided.

I can see the point. We all become victims of our emotions: If one is overwhelmed by deep romantic love or furious hatred, then one is bound to misperceive conditions by misattributing biased qualities to objects.

If one is able to disengage from these misperceptions, then reality loses these assigned qualities and becomes easier to handle.

So, mental practice, introspection, and cultivation of the mind are used to attain more objectivity. In addition, this "science of the mind" can serve as the basis of an ethical system. This seems to differ from viewpoints cultivated in Western societies where objectivity is thought to be attainable only from the third-person perspective by relying on criteria such as reproducibility, confirmation of predictions, and so on, and where ethics is not an integral part of scientific exploration, at least not in the natural sciences.

Matthieu: Provided we understand ethics as a science of happiness and suffering, not as dogma disconnected from lived experience.

Wolf: The premise is that mental practice leads to the construction of realistic models of oneself and the world. These novel insights, together with the effects of the practice, would then entrain changes in attitude, which, if shared by many in the long run, could improve the human condition.

Matthieu: This will arise from a way of being that has become free from those biases and mental entanglements, and therefore naturally expresses itself as altruism, compassion, and concern for others. This is in particular based on acknowledging that others want to be happy and avoid suffering, just as you do. Such a way of being will express itself spontaneously in ways that are beneficial to others. Your actions will be a spontaneous expression of your way of being.

Wolf: Provided everybody gets his model of the world right, this might work out.

Matthieu: If you maintain proper understanding or perspective, proper view, proper motivation, proper effort, and proper conduct, then it will certainly work in the best possible way. Even if life events and circumstances are unpredictable and beyond our control, we can always try to maintain our direction using the inner compass of right view and right motivation. This is the best way to achieve the goal of freedom from suffering for oneself and others.

Wolf: How can you harvest the wisdom that we believe is collectively gathered over generations and is codified in religious and legal systems—wisdom about how to behave that no single individual can gather in one lifetime? Certain attitudes may be beneficial for the individual's own life trajectory but may have long-lasting detrimental consequences for society that the individual will never experience. Such knowledge can only be harvested collectively across generations and cannot possibly be obtained by introspection alone.

Matthieu: Although you may not see the final outcome of your actions, as a fundamental principle, you can always check your motivation: is it a mostly selfish motivation based on exaggerated self-cherishing or a genuinely altruistic one? If you keep on generating such a motivation and then use the best of your knowledge, reasoning, and skills to act in that direction, then the effect is much more likely to be positive in the long run.

A correct understanding of reality leads to a correct mental attitude and moment-by-moment behavior that is attuned to that understanding. This in turns leads to a win-win situation of flourishing oneself while acting in a way that is also beneficial to others. Such an optimal way of being will have positive effects first in the family and then in the village or local community and gradually in society at large. As Gandhi said, "If we could change ourselves, the tendencies in the world would also change. As a man changes his own nature, so does the attitude of the world change towards him. ... We need not wait to see what others do."

Wolf: This is a profound insight on which probably all spiritual people would agree. However, in a highly interconnected societal system, it matters how others react to the transformation of individuals. Unless a substantial fraction of individuals follow the path of individual transformation, the danger remains that those clinging to power and selfishness will usurp the benevolence of a peaceful minority for their interests. We need normative systems that constrain the power and influence of defectors. Individual transformation and the regulation of social interactions will have to go hand in hand. Here we see the same complementarity of strategies that we encountered when we investigated differences between contemplative and natural science or between first- and third-person approaches toward a better understanding of the world and a betterment of the human condition. At some stage of our conversation, I hope we will explore to what extent convergence exists between the insights gained from contemplative science, the humanities, and the natural sciences. This comparison should be particularly interesting because the methods applied in the natural sciences are now also applied to investigate psychological phenomena accessible only through a first-person perspective, such as perceptions, feelings, emotions, social realities, and, last but not least, consciousness.

INVESTIGATING THE SELF

Is the self an entity that sits in the heart of our being or a command post in the brain? Or is it a continuum of experience that reflects a person's history? The Buddhist monk deconstructs the idea of the unitary, autonomous self, whereas the neuroscientist confirms that no cerebral zone takes on a central role in the brain. The idea of the self as conductor is a convenient illusion to function in existence. Is a strong "I" necessary for good mental health? Doesn't disposing with a blind belief in the self make us vulnerable? Would a transparent ego, on the contrary, make the soul stronger and favor inner confidence?

Matthieu: Let's come now to the root of misapprehending reality: grasping to the notion of there being a separate, autonomous self that would be the core of our being and would stand as the central command post of our experience. According to the Buddhist view, to postulate the existence of such a self-entity is a distortion of reality that feeds delusion and triggers all kinds of afflictive mental states.

Wolf: How can you reconcile this with the need to have a strong self?

Matthieu: It all depends what you mean by "strong self." A crucial difference exists between an inner confidence combined with a strong determination to achieve certain goals and the strong grasping to the belief that there is a central, distinct entity that constitutes the essence of our being. Inner strength does not come from having a reified ego and extreme self-centeredness but rather from inner freedom, which is quite different.

But let's first examine why we even entertain the feeling that we have an autonomous self. At any time, I feel that I exist, that I am cold or hot, hungry or replete. At every moment, the "I" represents the subjective, immediate component of my experience.

There is also the story of my life, which defines me as a person. This is the continuum of all that I have experienced through time. The "person" is the complex, dynamic story line of our stream of consciousness.

Wolf: This is in essence the biographic memory.

Matthieu: In a way, yes, but it is not only what we can actually remember. When you take a sample of water from a river, its quality and contents reflect the whole history of the river upstream from the location where you took that sample. Its quality depends on the soil and vegetation it has traveled through and its degree of pollution or purity.

These two aspects, the real-time "I" and the experiential continuity of the "person," enable us to function in this world. There is no problem with these two. But then we add something else: the concept of an autonomous self.

We know that our body and mind change at every moment. We are no longer restless children, and gradually we are becoming elderly. Our experience is changing and being enriched moment after moment. Yet we think that something within all this defines us now and has defined us throughout our life. We refer to this as "me," "self," or "ego." Not content to be a unique continuum of experience, we assume that, at the core of it all, there is a separate, unitary entity that is our true self, something like a boat that travels along the river of our experience.

Once we believe in such a self-entity, with which we identify, we want to protect it and fear its disappearance. This powerful attachment to the notion of self engenders the notion of "mine": "my" body, "my" name, "my" mind, "my" friends, and so on.

We cannot but conceive of this self as a distinct, unitary entity, and despite the fact that our body and mind undergo ceaseless transformations, we obstinately assign to this self the qualities of permanence, uniqueness, and autonomy. The main effect of this belief is not a genuine sense of confidence but, paradoxically, an increased vulnerability. First, by assuming that the ego is an autonomous, separate, unitary entity in the midst of our experience, we are basically at odds with reality. We only exist through interdependence, relations, mutual causality, and numberless causes and conditions. Our happiness matters of course, but it can only happen through and with the happiness of others. Furthermore, the self becomes a constant target for gain and loss, pleasure and pain, praise and criticism, and so on. We feel that the self must be protected and satisfied at all costs. We feel aversion to anything that threatens the self and attraction to whatever pleases and reinforces it. These two basic impulses of attraction and repulsion give birth to a whole sea of conflicting emotions—anger, craving, arrogance, jealousy, to name but a few—and, in the end, suffering.

Wolf: You seem to have already a rather sophisticated relationship with your "self." If you go around and ask people, "Who is your self?" they would probably just say, "It's me," and even seem puzzled at the question. They wouldn't take the stance of the observer that you are now taking in relation to yourself.

Matthieu: That's very true. As long as we don't examine the "self," we take it for granted and identify closely with it. But as soon as we start looking for and analyzing it, we realize how difficult it is to pinpoint anything as being the "self." It should be easy to see how much disturbance our grasping to this notion of a separate self brings to our lives. Yet we usually don't pay much attention to that question. Hence, it is difficult to get out of the vicious circle of suffering.

We can, however, help people realize intuitively the importance that the "I" takes in their life. If, for instance, I shout in front of a cliff, "Hey, Matthieu, you are such a damned stupid fellow!" when the echo of these words comes back to me, I laugh and don't get upset. But if someone next to me shouts the same words directly at me, with the same tone of voice and same intensity, I get annoyed. What's the difference? In the first case my ego was not targeted, but in the second case it was, which made the exact same words suddenly upsetting to hear.

Wolf: You were not upset when it was you who shouted the insults for the same reason that you can't tickle yourself: We perceive self-initiated actions in a fundamentally different way than actions initiated by others.

INVESTIGATING THE SELF

Matthieu: Because the concept of an independent "self" influences so much of our experience, we must examine it thoroughly. So how should we proceed?

Our body is just a temporary assemblage of bones and flesh. Our consciousness is a dynamic stream of ever-changing experiences. Our personal history is nothing but the memory of what is no more. Our name, to which we give such importance and to which we associate our reputation and social status, is nothing more than a grouping of letters, a label devoid of any essence. When I see or hear my name, Matthieu, my mind jumps. I think, "That's me!" But if I separate the letters that compose my name, M-A-T-T-H-I-E-U, I don't identify with them anymore. The idea of "my name" is just a mental fabrication. However much we explore our body, speech, and mind, we can never point a finger to any particular entity that could be the self. We have to conclude that the self is nothing more than a concept, convention, or designation attached to the combination of our body and stream of consciousness.

For such an investigation to make sense, we need to pursue it to its end. We need to conduct a thorough introspective investigation to determine the nature of the self. By doing so, we come to conclude that the self does not reside outside the body or in any specific part of the body. Nor is it something that permeates our entire body, like salt dissolved in water.

We might then think that the self is associated with consciousness. But consciousness is just a stream of experience: The past moment of consciousness is dead (only its memory and impact on the present moment remain), the future is not yet born even though we try to imagine it in the present, and the present is ungraspable. So there is no truly existing self, no soul, no ego, no *atman*—the essential self according to Hinduism—no personal, autonomous entity. There is only a stream of experience. Interestingly enough, this does not diminish us in any way; it merely frees us from serious delusion. That being said, it is perfectly fine and practical to consider a *conventional self* that is a label attached to our body and mind, in the same way that it makes sense to give a name to a river to help distinguish it from another river. The self does exist, but only in a conventional way, not as a truly existing, self-defined, separate entity. In short, the self is a practical, convenient illusion that allows us to define ourselves with regard to the rest of the world.

Wolf: True, inasmuch as you define the self as assigned to you by your social environment, as something attributed to you by others who perceive you as an intentional agent, an autonomous self, and inasmuch as you understand your self as an experience arising from the sum of your biographic memories and the awareness of being an embodied individual endowed with a mind.

It is true that you don't find the self or the mind linked to any discrete region in the brain, whereas there are centers responsible for biographic memories and body awareness.

Matthieu: There are more ways to show that the self can't be considered to be a circumscribed entity. When I say, "This is my body," the "me" becomes the owner of the body, not the body itself. But if someone pushes you, you complain, "He pushed me!" Now suddenly the self has become identified with the body. Then you go on, "He has hurt my feelings." Now you become again the owner of your feelings. Finally, you conclude, "I am upset." Back to the self being identified with the subject itself.

Let's imagine the self not as a localized entity but as something that pervades my whole body and mind. What happens to it if I lose both legs? In my mind, I am Matthieu without legs, but I'm still Matthieu. Even if I have poor self-image, my perception of there being a self deep within me has not been amputated to a third of its size. It has simply become a frustrated self, a depressed self, or maybe a courageous and resilient self.

Because we can't really find the self within the body, we turn to consciousness. However, unlike the idea of a constant self, conscious experience changes from moment to moment. So where is the self in all that? Nowhere.

Wolf: There are patients with complete amnesia who have a complete loss of episodic and biographic memory, such as the famous patient H.M., who had the temporal lobes of both hemispheres removed because of intractable epilepsy. H.M. lived only in the present but still had the notion of self. Thus, experiencing oneself as a product of one's individual history does not seem necessary for the constitution of the self. According to Brenda Miller, who followed H.M. for decades until he died in 2008, H.M. clearly experienced himself as a self, although he had no continuous biographical memory. He had, however, some recollections of memories dating from before surgery, old memories that he associated with "himself."

Matthieu: It would have been interesting to ask H.M. specific questions to get a better idea of the kind of representation of the self that he had. I would guess that the tendency to perceive a self entity associated with our being alive is quite instinctive and may not require one to have a lot of memories from the past. H.M. apparently related nicely to people. I met Brenda Milner and asked her about that. She confirmed that H.M. would react normally to someone calling his name and clearly had an image of himself. He would he react with satisfaction at being praised and became upset when someone criticized him.

Wolf: Apparently, he had a sense of humor and responded normally to praise and criticism. He did not understand questions referring to events that occurred after surgery, but his short-term memory span was sufficient to entertain conversations.

Matthieu: Did he have any social sense, for example, to distinguish between people who might have a "superior" or "inferior" social rank, such as the head of hospital versus the cleaning staff?

Wolf: This is not clear. He was nice to everybody but could recognize rank and familiarity, probably through unconscious processing of the relevant cues.

Matthieu: Did he expect to be treated in a certain way and become upset when not treated in that way?

Wolf: He got annoyed when asked to do something that he could not accomplish.

Matthieu: In short, would one sense that his immediate emotional reactions were influenced by a kind of self-centeredness or ego grasping? Was there any indication that his apparent general good mood was more eudemonic than hedonic?

Wolf: This issue is difficult because he was tested only post hoc and was unhealthy before due to the seizures. He certainly had his "ego," could be offended and flattered, and responded adequately to social interactions. Much of his "personality" was intact; he just could not consciously recall past events. To what extent his subconscious processing of sensory signals and subconscious access to past experience was still possible is of course difficult to clarify. I would assume that all procedural learning was still intact and got him through life rather well.

Matthieu: It would be fascinating indeed to know more in detail what kind of self someone like H.M. constructs in his mind. Personal history, the sense of having a self-image, the way we see ourselves, the way we would like others to perceive us, and the way we think others perceive us: All this could be drastically altered in someone like H.M. Yet there is perhaps no contradiction with what I mentioned earlier. Adding the concept of an "ego," an autonomous "self," may happen at a quite basic level when relating to the outer world. The question would be how much someone like H.M. would solidify this self-centeredness into a concept.

But to come back to our analysis of the self, in conclusion, the only thing we can say is, yes, there is a self, but it's simply a mental label for the stream of our experience, the association of our body and consciousness, which is made of parts and is ephemeral. There is nothing more than a conceptual, nominal self. Why should we be so obsessed with protecting and pleasing the self at any cost?

Wolf: Well, of course you want to protect yourself. You want to get through life without too much injury, and it is not a separate self that you want to protect, it is you in your entirety, it is you as a person. Even animals with no concept of self protect themselves, become aggressive when menaced, and show signs of comfort when attended to or caressed.

Matthieu: That's right. It is indeed natural and desirable to protect one's life, avoid suffering, and strive toward genuine happiness. What

I am referring to is the dysfunctional drive to protect the ego. In the example I gave of being upset and laughing at the same words, it is only the ego that aches in the first place and is amused in the second one. Your life isn't in danger. Yet it is from that exacerbated self-centeredness that most afflictive mental states come. When we entrench ourselves in self-centeredness, we also create a much deeper divide between us and the world.

Let me give you another example. Let's compare the stream of consciousness to the Rhine River. It has of course a whole history, but it is also changing all the time—"One can never step twice in the same river," said Heraclitus—and there is no such thing as a "Rhine" entity.

THE SELF EXISTS IN A CONVENTIONAL WAY

Wolf: Still, it is entirely legitimate to call that river the Rhine because many features of this river are constant and invariant despite the ever-changing properties of the water. Actually, this impermanence is a constant and constitutive feature of rivers.

Matthieu: Sure, and many features differentiate it from the Ganges: the landscapes along its banks, the quality and quantity of the water that flows along its stream, and so forth. Yet there isn't a separate entity that qualifies as the core of the Rhine's existence. "Rhine" is nothing more than a convenient designation attached to an ever-changing set of phenomena. To conceive the self in the same way is absolutely fine and does not prevent us from functioning in the world.

I remember a time in 2003 when His Holiness the Dalai Lama taught in Paris for a whole morning on the notion of no-self. At lunchtime, I mentioned to him, after looking at the written questions collected from the audience, that many people had difficulties in grasping this notion. They were asking, "If I have no self, how can I be responsible for my actions?" or "How can one speak of karma if there is no one to experience the results of past deeds?" And so on. His Holiness laughed and told me, "That is your fault. You did not translate properly. I never said that there is no self." He was teasing me, but what he meant and explained again in the afternoon is the fact that there is indeed a conventional, nominal self attached to our body and mind. That concept is fine and functional as long as we don't conceive of it as being a kind of central, autonomous, lasting entity that constitutes the heart of our being.

Wolf: I believe I am beginning to understand the nature of this misconceived self as you describe it. You consider it a projection, not an integral part of you that exists on its own. It is a mental construct detached from the roots of the person, and therefore needs continuous reassurance, reconfirmation, and efforts to make it appear as what you want it to be.

Matthieu: Well said. You fabricate an entity that you then install as the ruler of your world.

Wolf: Can we call this a projection?

Matthieu: Yes, it is a mental projection. The Buddhist analytical approach aims at deconstructing this mental fabrication through logical and experiential investigation and comes to the conclusion that we are not that imaginary entity with which we identify ourselves but rather a continuous, dynamic stream of experience.

Wolf: But you would not go as far as the philosopher of mind, Thomas Metzinger, who titled his book *Being No One*.[1]

Matthieu: We are certainly people, with a continuum of experience relating to a body and the outer world. Yet this continuum is in perpetual transformation, and we cannot find anywhere a singular, well-circumscribed entity that would be the "one," the concrete essence of our being. Metzinger also speaks of the self as an "ongoing process," of a "phenomenal self" that is not a discrete entity.

Wolf: I make a distinction between the concepts of self and self-centeredness. Self-centeredness has this exaggerated, egocentric connotation that we associate with egoism, an attitude that is not compatible with altruism, with compassion. However, the "self" is a term or concept that designates something that can be experienced, through either introspection or observation of others. One of these nontangible realities was brought into the world by human cognition and social interactions and affects our lives in the same way as belief and value systems or the notions of free will, autonomy, and responsibility. I identify you as yourself and can judge whether you have changed. I can say "You are no longer yourself" when you get angry or passionate, for example. But your identity would not be questioned.

THE SELF AND FREEDOM

Wolf: As concerns freedom, you know that I am not a defender of the concept of this unconditional free will that goes along with a dualist position and is impossible to reconcile with neurobiological evidence. I sympathize with the pessimistic philosopher Schopenhauer, who was clear in saying that you cannot will otherwise than your will, and you cannot change your will by mere will. At a later stage, we will engage in a discussion as to how free we are to consciously decide for or against something and what the neuronal constraints are.

For the moment, I would like to focus on the more sustained feeling of freedom, of being unconstrained, of being in harmony with oneself. Such feelings of freedom are experienced when there is congruence between subconscious dispositions and drives and the imperatives derived through rational analysis of the world. These are pleasant states in which one feels free and unconstrained by inner conflicts, possessive affects, or commands imposed by one's ego or outside forces.

The range of one's options can be significantly reduced by outer constraints. When these constraints are absent, one may feel free, but real freedom is when the various drives, desires, and constraints within oneself are in harmony with one another. We are programmed to strive for novelty, and at the same time we have a strong inclination for bonding and stability because, in such situations, there is no need for the intervention of the intentional self to settle conflicts by initiating a change. We are often confronted with incompatible drives. To bring the mind to equilibrium, to create the feeling of freedom, those internal competitions between incompatible desires have to be settled.

Matthieu: These internal conflicts are basically related to the two fundamental impulses of *attraction* to what is deemed pleasurable and *repulsion* toward the opposite.

Wolf: Yes. If those can be reconciled, if there is no conflict between self-imposed obligations and the results of rational deliberations, then one feels free and coherent, in which case the awareness of self may actually fade because one feels no constraining border. However, as soon as there are constraints, the self manifests itself as an agent whose freedom needs to be defended.

Matthieu: Inner conflicts are also unnecessarily created by an exacerbated sense of self-importance, as it becomes increasingly demanding. If you don't feel the need to defend the self because you have understood its illusory nature, then you will be much less prone to fear and inner conflicts. True freedom means to be free from the diktats of this self rather than following every single fanciful thought that comes to mind.

WEAK SELF, STRONG MIND

Wolf: But it also seems that many problems arise if one has a weak self that depends too much on others to define itself. Only then does one enter the vicious circle of wanting and repulsion.

Matthieu: There can be quite different reasons for this. Some people are tormented by the feeling that they are unworthy of being loved, they lack good qualities, and they are not made for happiness. These feelings are usually the result of scorn or repeated criticism and contempt by parents or relatives. Added to this is a feeling of guilt: Such people judge themselves responsible for the imperfections attributed to them. Besieged by these negative thoughts, they constantly blame themselves and feel cut off from other people. For these people to go from despair to the desire to recover in life, we must help them establish a warmer relationship with themselves and to feel compassion for their own suffering instead of judging themselves harshly. From there, they will also be able to improve their relations with others as well. The benefits of cultivating self-compassion have been clearly shown by researchers and therapists such as Paul Gilbert and Kristin Neff.[2]

In many other cases, what people usually call a "weak self" looks more like an insecure, capricious self, stemming from a confused mind that is never satisfied and always whines about its dissatisfaction. This is often the result of excessive rumination about oneself, of brooding "me, me, me" all the time, and being overly concerned by the slightest ups and downs in life. The illusory self wants to assert its existence by either being overdemanding or defining itself as victim. A person who is not preoccupied by self-image, self-assertion, and so on is actually much more confident, neither a narcissist nor a victim. A person with a transparent self is not vulnerable to pleasant and unpleasant circumstances, praise and criticism, good and bad image, and the like.

Wolf: Now, what about a strong ego? Would you equate that with a high level of self-centeredness?

Matthieu: I would not call it a strong ego but an inflated one. The only thing that is strong here is grasping. A so-called strong ego is in fact more vulnerable because, within its self-centered universe, everything becomes either a threat or an object of craving.

In addition, the stronger the ego, the larger the target you offer to the arrows of outer and inner disturbances. Praise preoccupies you as much as criticism because it inflates your ego further and makes you worry about losing your good reputation. When ego grasping dissolves, the target disappears and you stay at peace.

Wolf: What you are describing is in my view a narcissist's ego, which usually goes along with reduced confidence in oneself and hence with what I would call a weak or nonconsolidated ego. People with such personalities constantly need external support for their identification, and this, as you said, makes them vulnerable. We seem to have a problem of terminology here.

Matthieu: Yet studies have now shown that in fact narcissistic people do indeed have high self-esteem and are not just trying to compensate for concealed low self-esteem.[3] The ego can attain only a contrived confidence built on unreliable attributes such as power, success, beauty, and fame and on the image that we want to project onto others. The sense of security derived from that illusion is eminently fragile. When things change and the gap with reality becomes too wide, the ego becomes irritated or depressed, freezes up, or falters. Self-confidence collapses, and all that is left is frustration and suffering. The fall of Narcissus is a painful one.

Wolf: Fine, but most of the time we would say, the stronger the self, the more independent and autonomous you are; by being at peace with your self, the less you will be confused by misconceptions, the less you will suffer from egocentrism, and the more you will be able to develop empathy, generosity, and love toward others.

Matthieu: I think the main point here is to distinguish between a strong self and a strong mind. A strong *self* comes with excessive self-centeredness and a reified perception of a self-entity. A strong *mind* is a resilient mind, a free mind, a wise mind that is skillful in dealing with whatever comes one's way in life, a mind that does not feel insecure and is therefore open to others, a mind that is not swayed by anger, craving, envy, or other mental disturbances. All those qualities actually come from a reduced sense of selfhood. So we could say, in what may seem like a paradox, that the mind can only be strong when it does not fall under the empire of ego grasping. In short, the optimal situation would be weak self, strong mind.

In the same way, we should not confuse healthy autonomy or self-reliance based on inner freedom with grasping onto the notion of a reified self, which turns out to be the source of our vulnerability, chronic dissatisfaction, and demanding too much on others and the world. Again, when you speak about being independent, that is fine if you refer to the ability to stand on one's own feet and have the inner resources to deal with the up and downs of life. But this "independence" does not require conceiving of an independent self-entity. It is rather the opposite: By understanding the fundamental interdependence of oneself and others, of oneself and the world, we form the logical ground for developing altruistic love and compassion.

Let's not confuse self-grasping with self-confidence. Someone like the Dalai Lama, for instance, is deeply self-confident because he knows through his direct experience that there is no "ego" to be defended or promoted. So he laughs equally at the idea of those who say that he is a "living god" and at his "Chinese brothers and sisters" who say he is a "demon." The clearer your realization that the self has a merely conventional existence, the less vulnerable you will be and the greater inner freedom you will gain.

Wolf: If there is no projection of your self, then there is no self to be attacked, I agree. But it seems to me that I can still insult or attack you, and you would be offended and defend yourself.

Matthieu: The best defense is for me not to be affected at all. It does not mean that I am dull or stupid, but that your actions and words simply have no impact on me. Someone may wield a sword through space with great anger, but it can do nothing to space. Someone may throw dust and colored powders at space, but they will just fall back on his own head. When ego grasping does not offer an easy target for insult and praise, instead of being upset, you will just laugh, as the Dalai Lama does or as an old man would watching children at play: He clearly sees all that is happening, but he does not get upset as the children often do when one side wins or loses. This vastness of mind, this freedom, is a sign of the inner accomplishment achieved through meditative practice.

Someone who can rest in a natural, unperturbed, selfless state of mind is not at all indifferent to others and aloof from the outer world but can rely on readily available inner resources.

Wolf: I understand. What you call a strong mind that comes with a transparent self is probably equivalent with my strong or well-consolidated self that requires little attention because it rests in itself and requires no affirmation, whereas the self-centered self to which you ascribe the many negative connotations would be my weak or insecure, egocentric, egoistic, or even narcissistic self that requires permanent confirmation of its shallow existence.

Matthieu: In one Buddhist text, it says that in the beginning, to recognize the effects of self-clinging clearly, you need to somehow let the self manifest in its full force and observe what it does in your mind. Then you have to investigate its nature and, having recognized its conceptual nature, deconstruct it. In other words, you should not ignore it but observe how it works and then transform it into a state of freedom. At this point, genuine confidence will arise.

Wolf: It is not easy to intuitively grasp this. I would assume that once your confidence is no longer dependent on external reinforcement, you are simply no longer concerned with it and can let loose without being lost. This would seem to lead to the goal of healthy maturation toward a self-reliant, independent adult personality. Children are still lacking this sovereign ego and therefore depend much more on external reinforcement.

Matthieu: Sure. You may also say that a person whom we call free is thus called because that person is free from all kind of fetters, whether the inner fetters of clinging or the outer fetters that come from unfavorable circumstances. Self-reliance comes with freedom, not with an emperor-like, overarching ego.

EGO AND EGOLESSNESS

Matthieu: People with a transparent self are much more pleasant to be with. They also feel much more connected to others because many of our problems come from creating an artificial gap between self and others as being fundamentally separate entities. By doing so, the self negates its interdependence with the world and wants to confine itself within the bubble of the ego. The French existentialist writer Jean-Paul Sartre wrote, "Hell is the other." I would rather say, "Hell is the self." Not the functional conventional self, but the dysfunctional superimposed self that we take as real and that we let rule our mind.

Our mutual friend Paul Ekman, one of the most eminent specialists in the science of emotion, has been reflecting on what he calls "people gifted with exceptionally human qualities." Among the most remarkable traits he has noted among such people are "an impression of kindness, a way of being that others can sense and appreciate, and, unlike so many charismatic charlatans, perfect harmony between their private and public lives." Above all, notes Ekman, they exhibit "an absence of ego. These people inspire others by how little they make of their status, their fame—in short, their self.[4] Ekman also stresses how "people instinctively want to be in their company and how, even if they can't always explain why, they find their presence enriching." Egoless people, like the Dalai Lama, are incredibly strong, and their inner confidence is like an unshakable mountain. That is not the case with people who are full of themselves. The least we can agree on is that they are usually not very inspiring to be with.

Wolf: My personal impression is that the Dalai Lama is like a rock in the moving sea: He is sensitive, he is reachable, but he does not depend on his self-esteem or the acclaim of others, nor does he feel hurt when somebody says, "I don't like you." However, for me this is a signature of a strong personality, of self-confidence in the positive sense, of a steady, non-narcissistic self. He is someone who clearly has no ego problems because he rests confidently in himself. Self-centeredness, dependence on acclaim, vulnerability in the face of criticism, and so on—I would equate these signs of reduced self-confidence with a weak, nonconsolidated ego, an ego that has not found its proper coordinate system.

Could you define for me more precisely this "ego bubble"? Is it something like an artificial sense of being at home within one's self-contained self?

Matthieu: You can only truly be at home within the freedom of pure awareness, not within the bubble of self-grasping. The ego bubble is a narrow mental space in which everything gravitates around the "I." You actually form that bubble with the illusory hope that it will be easier to protect yourself within a confined space that constitutes a kind of refuge for the ego. In fact, you have built an inner jail in which you are at the mercy of endless thoughts, hopes, and fears that keep swirling around within that bubble. This feeds an exacerbated feeling of self-importance and self-centeredness that thinks of nothing but achieving its immediate satisfaction, with little concern for others and the world, except for the ways in which it might use them or be affected by them.

The problem is that within such a bubble, everything gets blown out of proportion. The smallest upheaval will upset you greatly. Self-centered thoughts keep bouncing back and forth on the bubble's illusory walls. You lack inner space.

If you burst the ego bubble and the small space of your grasping mind expands into the vast expanse of pure awareness, then the same events that were so upsetting before now seem rather inconsequential.

Wolf: Because we seem to have differing concepts concerning weak and strong egos, can't we find another word for the trait that imprisons one in the ego bubble?

Matthieu: Would "self-centeredness" seem appropriate?

Wolf: I am happy with that. Self-centeredness catches the essential traits and has the right kind of connotation.

Matthieu: I will add that often people who are engrossed in self-centeredness want you to experience the world the same way that they do, lest they feel rejected. For instance, some may entertain pessimistic views about the world and other people, which lead them to distrust others. To feel appreciated, they want you to enter their ego bubble, adopt the same attitude as them, and function just like them. We could be quite prepared to earnestly take their viewpoint into consideration and be open to their way of being, but we can't adopt their way of thinking and worldview just to make them happy.

THE SCOURGE OF RUMINATION

Wolf: You suggested as one good contemplative practice the deconstruction of an "inflated" ego whose constitutive property is grasping and integration of this ego in a vast network of relations. I propose to now investigate how this can be related to the goals of conventional education or psychotherapy. I believe that psychoanalysis also tries to create an integrated self, but the procedure differs radically from contemplative strategies. It emphasizes the role of the self, encourages the self to become the judge, and, contrary to meditation, encourages rumination, the exploration of conflicts.

Matthieu: I am far from being an expert, but a number of people whom I know well and who have gone through years of psychoanalysis followed by years of Buddhist practice trying to achieve that sense of confidence born from inner freedom reported to me that they experienced a clear difference between the two approaches. One of them told me, "In psychoanalysis, it is always me, me, me, my dreams, my feelings, my fears," with a lot of rumination about the past and anticipation about the future.

Other people are also considered, not so much in themselves but through the filter of self-centeredness.

Although the psychoanalytic approach might help one develop an acute perception of mental constructs, it remains enmeshed in a swamp of ruminations about the past. It is like trying to find some sort of normalcy within the ego bubble instead of breaking free from that bubble. For a Buddhist practitioner, it sounds like a really bad idea to stabilize the ego and make a pact with it. It almost resembles Stockholm syndrome— people who end up feeling some kind of closeness and even sympathy with those who have kidnapped and abused them.

In Buddhist practice, one wants to break free from these entanglements, not just come to terms with them. The limpid awareness of the present moment is a complete freedom from ego grasping and rumination.

Wolf: So rumination and the analysis of conflicts would be the opposite of what you do during meditation?

Matthieu: Absolutely. Rumination is the scourge of meditative practice and inner freedom. Now, rumination should not be confused with analytical meditations, which will, for instance, deconstruct the concept of an independent self. Rumination is also different from the vigilant observation of your states of mind that will allow you to recognize the arising of an afflictive emotion and defuse the chain reaction that usually occurs. In fact, quite a few similarities exist between the Buddhist way and the methods used by cognitive and behavioral therapy to spot emotions and mental constructs that have caused our present afflicted mental states and help one to see how unrealistic these mental fabrications are.

Rumination has also been shown to be chronic and excessive in depressed people. One of the outcomes of mindfulness-based cognitive therapy is to distance oneself from those ruminations through mindfulness meditation.

Wolf: This idea is interesting. We seem to have two techniques, one ancient, the other quite recent, that both have as their goal the betterment of the human condition and the stabilization of the self, but they follow entirely different strategies. At first sight, meditation, too, may seem like a rather self-centered practice. One sits in isolation, is not communicating, and is fully occupied with oneself. Can meditation not also degenerate into solitary, selfish rumination?

Matthieu: It could, but that would be considered a deviation. As for proper meditation, how can it be considered a selfish endeavor when one of its main goals is to get rid of selfishness? Rumination is a troublemaker because it feeds endless chains of thought that keep people exclusively preoccupied with themselves. This is the opposite of remaining in the freshness of the present moment, in which past thoughts are gone and future thoughts have not yet arisen. You simply remain in pure awareness. Whatever thoughts might arise, you let them go without leaving a trace. This is freedom. In addition to this, you will also cultivate altruistic love and compassion. So, when you return to the world, you will be much better equipped to put yourself to the service of others.

Wolf: It would be highly desirable if one could really go through life in that relaxed way, if a little bit of what one experiences during meditation could be preserved in daily life, in situations where one has to take on responsibility and is confronted with incompatible conditions.

Matthieu: Of course, that's the whole point. The constructive effects of meditation must be maintained during the postmeditation period.

Wolf: It would indeed be wonderful if this were possible, if meditation were a technique to escape from the vicious circles of negative thoughts, distrust, revenge, and deception because these mental fabrications are contagious across the members of a social group. Once one begins to take an eye for an eye, a tooth for a tooth—

Matthieu: As Gandhi said, if we keep on doing so, the whole world will become toothless and blind. The contemplative methods offered here are created precisely to escape the vicious circle of negative thoughts.

Wolf: Weak personalities, your strong ego-grasping characters, continuously seek affirmative interactions to reassure their existence. It's contagious. Do you have any proof that meditation can help to break this vicious circle and immunize the practitioners against such deeply rooted behaviors?

Matthieu: We say that the signs of a successful meditation practice are a well-tamed mind, a vanishing of afflictive mental states, and a conduct that is in harmony with the inner qualities one has endeavored to cultivate. If meditation were just about feeling good for a while, relaxing and emptying your mind within yet another bubble of artificial tranquility obtained by removing oneself from disturbing circumstances, that would be pretty useless because as soon as one would be confronted with adverse circumstances or inner conflicts, one would, as before, powerlessly fall prey to their effects. So meditation practice has to translate into real, gradual, lasting change in the way you experience your own inner life and the outer world. I must say that many of the experienced meditators I have met in my life do display such qualities. Otherwise, all this would just be a plain waste of time.

WHO'S IN CHARGE HERE?

Matthieu: You once mentioned that the structure and functionality of the brain is more attuned to the Eastern idea of a self—that is, a construct that is the result of many interdependent factors or the emergent property of a network—than to the Western idea of a well-defined central post of command.

Wolf: There is indeed a striking discrepancy between our Western intuition about the brain's organization and scientific evidence. Most of the occidental philosophical positions are anchored in the conviction that the brain possesses a singular center that coordinates all neuronal functions. This would be the place where all sensory signals converge to be interpreted in a coherent way, where decisions are reached, values assigned, plans formulated, and responses programmed. Finally, this would be the place where the intentional, autonomous self has its seat.

   In contrast to this intuition, which has dominated occidental philosophies and belief systems and nurtured the concept of ontological dualism, neurobiological evidence traces a radically different picture. There is no Cartesian center in the brain. Rather, we face a highly distributed system that consists of multiple interconnected modules that operate in parallel, each devoted to specific cognitive or executive functions. These subsystems cooperate in ever-changing constellations depending on the tasks that are to be accomplished, and this dynamic coordination is achieved through self-organizing interactions within the networks rather than through top-down orchestration by a superordinate command center. These distributed but coordinated processes generate highly complex, spatially and temporally structured patterns of activity that are the correlates of perceptions, decisions, thoughts, plans, feelings, beliefs, intentions, and so on.

Matthieu: If there is no such central command, then how did the notion arise that we have of a self-entity, and why is this helpful in evolutionary terms?

Wolf: This question is closely related to the question of why we have the impression that our will is free of the constraints of nature even though we know that our decisions are the consequence of neuronal interactions that follow the known laws of nature. There is of course some noise in this complex system, but on average it works reliably according to the laws of causality. This is fortunate because otherwise it could not adapt itself to the world, could not make "correct" predictions, and could not respond adequately to the ever-changing conditions that organisms have to cope with to survive. The problem is, we have no sense faculty that can detect the processes within our brain that prepare our perceptions, decisions, and actions. We are only aware of the consequences of these hidden neuronal processes.

We have the same problem with the concept of the internal mover, the observer, or the central agent that we associate with the ego. We perceive the respective other as a unitary intentional agent and attribute the same to ourselves, without being aware of our underlying neuronal processes. If anything, our intuition suggests that our self or our mind is somehow at the origin of our thoughts, plans, and actions. It is only through neuroscientific exploration that we discover there is no singular locus in the brain where this intentional agent could be located. All we observe are ever-changing dynamic states of an extremely complex network of densely connected neurons that manifest themselves in observable behavior and subjective experiences.

Matthieu: So the problem might be that we feel there *should be* a singular agent to explain the way we think and act, and we are puzzled when we don't find one.

Wolf: It is exactly because of the difficulty in imagining that these immaterial phenomena such as perceptions, agency, and emotions could result from material processes that ontological dualism, a strict ontological separation between mind and matter, has been postulated over and again throughout history. If one considers brains as simple machines that are composed of matter that obeys the laws of nature, one is indeed bound to postulate an immaterial, independent mover that is endowed with all the properties that we associate with the self. However, scientific analysis of the brain contradicts such simplistic views.

The brain is a complex system with nonlinear, self-organizing dynamics. It has been adapted by evolution, education, and experience to pursue certain goals and actually can accomplish all the functions we attribute to the self—at least this position is held by the majority of cognitive neuroscientists today. Such complex systems evolve along trajectories that cannot be predicted, although the trajectories can be accounted for in retrospect—just as is the case with evolution. Thus, nonlinear, self-organizing systems are creative and capable of behavior that a naïve observer would characterize as intentional, goal directed, and sensible. We tend to deny that such properties can emerge from the dynamics of our brains because we have no intuition for the complexity and nonlinearity of this organ. We believe that it follows the seemingly simple rules that govern the small fraction of processes in nature that we can perceive with the particular senses with which evolution has endowed us. This false belief about the organization of our brain makes it imperative to postulate a homunculus that rules in our head and endows us with all the marvels that we attribute to the self.

Matthieu: This is probably also due to the fact that we need to translate a complex process into something simpler and that we feel more comfortable imagining that there is a unitary entity that is in charge. The problems start when you conceptualize this process as being a truly existent, distinct entity.[5]

Wolf: This is the same problem as with the concept of God. One wants to explain a host of phenomena that one cannot account for with the cognitive tools at hand, and so one invents an agent that throws the lightning, generates the thunder, and blows the storm.

Matthieu: So the reified "self" is our homemade "God."

Wolf: Yes, in a sense. One invents an intentional, independent agent that exists at a different ontological level but can influence the world in which we exist. In effect, these concepts, these constructions, these projections—we could even call them social realities because most of them are the product of social interactions, of interpersonal discourse—have a huge impact. They act on us, make us build cathedrals, and tell us when to feel guilty and when to help our neighbors. Also, by endowing them with power, a quality we assign to gods, we delegate responsibility; we attribute to them the role of a shepherd, judge, or ruler and thereby make them responsible for our happiness and misery. Finally, they can be assigned an absolute authority to ensure obedience to rules that have been identified as useful by collective experience such as the Ten Commandments. To transcend the authority renders it immune to relativism and makes discussions obsolete because there will be no answer.

Matthieu: Mental constructs like the self or ego may provide simplifying explanations, but at some point they stop being helpful because they do not reflect reality. Conversely, if instead of perceiving the ego as an inner lord we see it as an interdependent stream of dynamic experience, it might be a bit uncomfortable at first, but it helps to free us from suffering, for the very reason that it offers a vision more attuned to reality.

Wolf: Is there a price you pay for that step?

Matthieu: I don't see any side effect to this recognition. All that you gain is inner freedom and a genuine sense of confidence and happiness because the ego's grasping is truly a magnet for suffering. That's what Buddhism says about it.

Wolf: I guess we should come—should we not?—to an end.

Matthieu: Sure. It was a wonderful day.

FREE WILL, RESPONSIBILITY, AND JUSTICE

THE PROCESS OF DECISION MAKING

Does free will exist? If all our decisions are made by neuronal processes of which we are only partially conscious, are we really responsible for our actions? To what extent are unconscious mechanisms influenced by all the experiences we live through? Can training the mind modify the content of unconscious processes and how they work? What are the repercussions on our way of seeing personal responsibility, the ideas of good and evil, punishment, rehabilitation and forgiveness? In the end, can free will be proven?

Wolf: Should we now examine the notion of free will? I propose to first give you the arguments that I presented at an international conference of philosophy a couple of years ago. My motivation was the feeling that our legal system puts too much responsibility on the shoulders of forensic psychiatrists. Unfortunately, at that time, the discussion quickly became polemical because the media stoked the debate with a false conclusion: If there is no free will, there is no guilt, and hence no justification for punishment.

Matthieu: The forensic psychiatrist is the person who examines the circumstances of the crime and tries to guess the motives of the criminal?

Wolf: No, the forensic psychiatrist is the expert who is asked by the court of justice to judge whether the accused subject is fully responsible for his or her actions or whether there are extenuating circumstances. Eventually, this expert decides whether the accused should go to jail or be considered mentally ill and transferred to a psychiatric facility.

For a neurobiologist, it is obvious that everything a person does is prepared by the neuronal processes that take place in his brain. As far as we know, these processes follow the laws of nature, including the principle of causality. Otherwise organisms would be unable to establish consistent relations between environmental conditions and adapted behavioral responses. If organisms responded in a random way to the challenges of the world, they would not survive. A nondeterministic brain would make you sometimes run away from a tiger and at other times just stay in front of it.

Matthieu: Or try to pat the tiger on the head...

Wolf: The probability of survival and reproduction would be rather low. Organisms endowed with such unreliable brains are unlikely to be among our ancestors. Neurobiology posits that all mental processes—including those that appear to be remote from material processes such as having feelings, reaching decisions, planning, perceiving, and being conscious—follow neuronal processes rather than initiate them. In such a framework of explanations, it is inconceivable that an immaterial mental entity makes the neuronal networks execute what this entity wants them to do to generate an action. Neurobiology takes the strong position, and I think rightly so, that whatever enters our consciousness is the consequence of neuronal activity in the large number of centers in the brain that need to cooperate to produce the specific states that we experience as perceptions, decisions, feelings, judgments, or will. Thus, from this perspective, all mental phenomena are the consequence and *not* the cause of neuronal processes.

Matthieu: But isn't it true that all you can really speak about are correlations between neuronal processes and mental events? The question of causality does not appear to have been solved so far. I may just as well argue that directly training the mind affects neuroplasticity. So there seems to be a two-way, mutual causation.

Wolf: We have more than just correlational evidence! Specific lesions lead to a specific loss of function, and electrical or pharmacological stimulation of particular brain systems induces specific mental phenomena and feelings of well-being or fear, and modifies perceptions and actions in predictable ways. If you train your mind, there must be a motivation for you to do so. This motivation is a reflection of a particular neuronal state (i.e., specific neuronal activity patterns that eventually generate the motivation patterns that make you sit and contemplate). These motivation patterns may be induced by the instructions of a teacher, which are then translated into neuronal activity patterns. Alternatively, particular internally generated brain states may give rise to the wish to contemplate. These could be memories of previously experienced beneficial effects or reports of others who recommended mental training. The actual trigger to contemplate might be an unresolved conflict or a solution for filling some spare time ahead. Both cognitive conditions will be associated with specific neuronal activity patterns. Thus, once sufficient motivation has built up, you will sit and engage in mental training, which is again associated with specific activation patterns. If sustained over a sufficiently long period of time, these patterns in turn will induce changes in the couplings among neurons and thereby generate long-term modifications of brain functions—just as training a movement will change the brain architectures responsible for the generation of this movement.

We may ask, what about arguments that occur to us or that are put forward by others that obviously influence our decisions and actions? Like any other mental phenomenon, they are the products of neuronal processes, and these processes will in turn influence and potentially change those neuronal activation patterns that underlie the subsequent decision or action. In the case of one's own arguments, the information substantiating the argument is derived from diverse neuronal sources— remembered experiences, stored moral values, emotional dispositions, and the actual perceived context.

The arguments of others also have a neuronal correlate in the brain of the receiver. A verbal argument is translated by the ear into neuronal activity; this is decoded semantically in the speech centers of the brain, and the resulting neuronal activation patterns then impact other brain centers and ultimately those that prepare the decision. So far, all data support this view, and there is no evidence that would force one to search for alternative explanations.

Matthieu: If I pursue your line of reasoning, I might then say that your own conviction, your present mental state, is the product of neuronal activity, which is in turn determined by multiple factors: the genetically specified architecture of the brain, the epigenetic modifications of this architecture caused by experience, and the actual context of a given experience or circumstance. Philosopher of science Michel Bitbol reminded me that phenomenologist Edmund Husserl argued that if we were to consider that "logic" is a mere product of human evolution and of the human brain, then simple principles of logic could not have any universal value. Understanding logic certainly depends on the presence of a conscience, but the proposition "if A is greater than B and B greater than C, this implies that A is greater than C" remains valid irrespective of the type of consciousness with which one is dealing.

Wolf: I see no contradiction here and agree with Husserl. All evidence indicates that our cognition depends on constructivistic processes. We already discussed this point. The a priori knowledge and computational algorithms required for this construction of experiences reside in the specific architecture of the brain. Because this architecture is the result of genetic and epigenetic adaptation to the world accessible to our senses, it is likely that our perceptions and way of making inferences are idiosyncratic and not generalizable. It is true that we are caught in an epistemic circle. Our brains and hence our cognition have become adapted to the small niche of the world in which life evolved. Within this tiny niche of the universe, only those variables have guided the adaptation process of our cognitive systems that can be encoded by our sense organs, which are highly selective and sensitive to only a narrow range of physical-chemical signals. Hence, we use a cognitive instrument that has been tuned to deal with a narrow segment of the world to "understand" the entirety of the world. We extrapolate from the dimensions of the universe to which we have been adapted to those dimensions to which we are not. Even worse, our cognitive instruments have not been optimized by evolution to analyze the hypothetical "true nature" behind phenomena, but to extract and interpret information that is important for the survival and reproduction of the organism. This requires heuristics that are different from the strategies necessary to obtain information about the true nature of things.

But let me return to the question of free will and the neuronal underpinnings of decision making. In addition to the evidence that decisions are prepared by neuronal processes that obey the laws of nature, we are usually aware of only a small fraction of the causes that determine our decisions. Using noninvasive imaging techniques, it has been consistently documented that subjects often become aware of the outcome of a decision seconds after the moment at which the analysis of their brain's activity allows an outside observer to predict what the decision will be. Thus, neuronal activity in the decision networks can converge toward a result—the decision—before the subjects become aware of having decided.

Matthieu: How many seconds?

Wolf: Up to 10 or 15.[1]

Matthieu: How do you know when the subject becomes aware of having reached a decision?

Wolf: Subjects are asked to decide at will when and with which hand they want to press a response key and to signal the outcome of their decision by executing the respective motor movement. So, the subject decides at a particular moment in time to press with the right hand. Subtracting the time required for the programming and execution of the motor act from the moment of pressing the button yields the time at which the subject became aware of having decided.

Matthieu: But then, what about the secondary intention to press the button to signal the intention to use the right or left hand? It seems that two intentional processes are involved here.

Wolf: The main point is that once subjects have reached a decision, once they make up their mind to respond with the right hand, they press the button. The registration of neuronal activity reveals, however, that the neuronal processes preparing this decision have started quite some time before. Complementary evidence comes from studies in which subjects execute actions in response to commands given in a way that prevents subjects from perceiving them consciously. In other words, subjects respond but are unaware of having followed an instruction. This can easily be achieved with split-brain patients. These patients have had the connections between the two brain hemispheres severed to control the spread of epileptic seizures. If a stimulus is presented to the nondominant hemisphere, which is less speech competent, patients remain unaware of having been exposed to a stimulus, but the nondominant hemisphere nevertheless processes and responds to the stimulus. A similar dissociation between awareness and stimulation can also be induced in healthy subjects. In this case, instructions are manipulated such that they remain below the threshold of conscious awareness (i.e., the stimuli remain subliminal).

For example, this can be achieved by "masking." If a stimulus, in this case a written instruction, is presented briefly and immediately followed by a high-contrast pattern, then the instruction passes "unseen" but may still be processed by the brain and solicit the appropriate action. Another possibility is to divert the subject's attention, a strategy commonly applied by magicians to render actions invisible. Once subjects execute responses to the "unperceived" instructions, to instructions they were not aware of, of course they become aware of their actions, but they experience them as the result of their own intention. If one asks the subjects, "Why did you do this?", they will respond in an intentional format: "I did it because I wanted to." Then they give an invented reason, convinced that it was their intention to initiate the action. This is a prime example of the illusory self-attribution of agency.

We have a need to find reasons for whatever we do, and when we have no access to the actual reason for our action, when the motivation is subconscious or our attention has been diverted in the moment of receiving the stimulus that initiated action preparation, we invent a reason and believe in it, without being aware of having invented it at all.

Matthieu: But why should the brain need reasons? Only "you," as an integral person, need reasons to be embedded in a self-narrative. Causal relations between brain processes are not reasons. Having a reason to do something implies a notion of goal, of being oriented toward a particular end, something that is not found in a mere succession of causal events.

Wolf: Persons become aware of their actions unless their attention is diverted, and they feel the urge to explain why they did what they did because they assume that all action has a cause and wish to stay coherent. If their action is triggered by a cause that they are not aware of, then they might admit that they don't know why they did it, but with all likelihood they will invent a reason post hoc. At the neuronal level, there is of course a seamless sequence of activation patterns whose succession obeys the principle of causality. Given all the neurobiological evidence on firm and causal relations between brain processes and behavior, the assumption appears untenable that a person, in the moment of deciding, could have decided otherwise. However, *this is what our legal system assumes:* The delinquent could have acted otherwise and because he did not, he is guilty and should be punished.

Interestingly, the question of free will and free decisions is discussed mainly in the context of decisions that are considered to be the result of conscious deliberations and reasoning. These might be moral arguments, stored in and retrieved from memory, arguments concerning the consequences of an act, beneficial or harmful, or arguments that have been heard recently. If there is sufficient time to weigh those arguments according to the discursive rules and value systems that are accepted in one's respective society, and if the ability to deliberate is unconstrained (i.e., if consciousness is undisturbed), it is assumed that one is absolutely free to choose between future acts, including the option to avoid deciding altogether.

The "deliberator" is, however, a neuronal network, and the outcome of the deliberation, the decision, is the consequence of a neuronal process that is in turn determined by the sequence of immediately preceding processes. Thus, the outcome of this process depends on all the variables that have shaped the functional architecture of the brain in the past: genetic predispositions, epigenetic effects of early imprinting, the sum of past experiences, and the present constellation of external stimuli. In brief, a pending decision is influenced by all the variables that determine how a particular brain is programmed and all the influences that act on the brain in the moment of making the decision.

Matthieu: But the brain activity that begins 10 seconds before the action was also influenced by countless conscious and unconscious events that preceded this current decision-making event. It seems to me that these data simply show that some brain events are associated with conscious thoughts and intentions and some are unconscious processes. Both precede and influence our actions. So the experiments that you report simply include unconscious processes in the picture. In fact all you are saying boils down to acknowledging the validity of the laws of cause and effect. But many other factors than unconscious processes can be potentially involved in the web of causation.

Also, when you say that the deliberator is a neuronal network, you may come to think, "It's not me who made that decision—it's my neuronal networks." In that way, you are dissociating yourself from your own actions, and you thus become unable to take responsibility for them at the level of the first person ("I am responsible for what I did"). This position is not neutral because it can further influence our decision making and behavior. It has been shown, for instance, that people who read a text claiming that all our behavior is fully determined by the brain behave quite differently than people who read a text asserting the existence of free will.[2] Interestingly, the people who were made to believe in free will behaved in more righteous ways than those who were made to believe in brain determinism. The latter were more prone to deviate from moral standards and to cheat probably because they felt that, after all, they were not really responsible for such behavior.

As for the experiment you mentioned about choosing a hand to press a button, our mutual friend Richie Davidson suggested that it would be interesting to see whether long-term practitioners remaining in open presence meditation could be aware of the decision to use their right hand earlier than naïve subjects, and even to see whether they could suddenly alter the process and switch to the left hand, even though brain processes would have predicted that they would surely lift the right hand.

In essence, as Kathinka Evers, a specialist of neuroethics, points out, even if conscious decisions are preceded by an unconscious neurological preparation, this does not mean that consciousness is not instrumental during prior conscious steps that have influenced the content of the unconscious processes. This means that we do have some control over unconscious processes through the content of preceding conscious ones. This also implies that we have some responsibility for the content of our unconscious because conscious and unconscious phenomena keep on shaping each other in an intricate web of mutual causation.

When some heroic person jumps into icy waters to save someone who is drowning, often when things are over and others congratulate and praise him, he says, "What I did is normal. I had to do it. I had no choice but to help." It is not really that he did not have any choice. Rather it is just a way of saying that for him the choice was so obvious that the decision to jump arose in a split second. So when things happen quickly, the way one acts spontaneously is a reflection of what we are: more or less altruistic and more or less courageous. But the way we are is the result of many conscious moments during which our mind turned more or less to altruistic ways of thinking, gradually building up an altruistic frame of mind and way of being. Therefore, even if during the few tenths of a second preceding a decision there has been unconscious processing in the brain, *the final decision is essentially the culminating point of a life-long experience.*

This also means that through mind training, we can fashion our conscious and unconscious processes, our ways of thinking, emotions, moods, and eventually our habitual tendencies. In particular, we are responsible for taking this process in the right direction and cultivating a moral, constructive way of being rather than pursuing unethical, harmful behavior.

Wolf: You are absolutely right. Conscious recollection of experiences and conscious deliberations do of course influence and eventually may also modify subconscious heuristics through mindfulness training, but one needs to keep in mind that these "conscious" processes are also brought about by neuronal interactions. This brings us back to the simple and undisputed notion that neuronal states influence subsequent neuronal states. There is no such thing as "consciousness" without a corresponding neuronal substrate.

Matthieu: But aren't you are jumping a bit fast to a conclusion? Most neuroscientists certainly think that way, but it would be an exaggeration to say that there is final, irrefutable proof of that. In fact, what you are describing is a causal process, which is absolutely fine. But are we sure that we are including all the possible causes that might influence our thoughts and decisions? If there were such a thing as a consciousness able to have top-down causal influence on the brain, then that would also be part of a causal process and would not imply something "spooky" or an exception to the law of causality.

Wolf: We will come back to the question of mental causation, but I wish to first conclude the point I wanted to make. The way we decide, the way our neuronal machinery converges toward a decision, depends on all the variables that influence the dynamic state of the brain in the moment of the decision. Several factors have shaped the functional architecture of our brains—genes, developmental processes, education, experiences—as well as influences from the recent past—arguments, context, emotional dispositions, and countless others. In principle, any past experience one is aware of can be taken into account in conscious deliberations. However, many experiences do not reach the level of conscious awareness and hence cannot figure as arguments in conscious deliberations. Still, they will influence the outcome of decisions as subconscious motives and heuristics. Actually—as mentioned before—only a small fraction of the many variables that intervene in decisions can be subject to conscious deliberations. We have limited conscious recollection, if any, of the genetic and epigenetic factors that shaped our brain architectures and, as a consequence, of our individual behavioral dispositions. The same holds for the many implicit motives that determine what we do next.

Matthieu: You say that we have limited conscious recollection of the factors that shaped our brain architecture. I would also say that most people have limited awareness of their own consciousness, of the tiny mental processes that keep occurring in the mind, as well as little, if no, perception of the pure awareness that is always present beneath mental constructs. While focusing on neurons and brain structure, we neglect to experience our own consciousness in the present moment, which could give us valuable insights into the nature of consciousness.

Wolf: I do not think that reflections on neurons and brain structure interfere with experiences of being conscious. What you allude to here is maybe a form of meta-consciousness, the ability to be aware of being aware. To cultivate this meta-awareness surely requires one to take a step back and escape from the hamster wheel, but why should reflections about underlying neuronal processes distract one from cultivating meta-awareness?

Coming back to the motives that determine our decisions, I find it particularly worrying that we may at times even be unable to retrieve from our stored knowledge the arguments that would prevent us from reaching a maladapted decision, if we could only access them during our deliberations.

Matthieu: It evokes what Paul Ekman called "the refractory period." When you are angry, you don't register or bring to mind anything that does not justify your anger, such as the positive qualities of the person with whom you're upset. I still think we should take a more holistic view and include the long-term influence of our past states of consciousness in our examination of decision making and free will.

Wolf: Indeed, the arguments that appear to consciousness are often subject to a selection process that depends on subconscious motifs and is exempt from volitional control.

Matthieu: So you would not be able to consider carefully every possible argument.

Wolf: Because of the limited capacity of the workspace of consciousness, at every instance, only a selected set of arguments are available for conscious deliberation, and the way in which these arguments are weighed and combined of course again depends on the architecture of the brain and its actual dynamic state. The former differs from person to person and the latter changes from instance to instance depending on the context.

Moreover, because the capacity of working memory is limited and shows interindividual variability, the number of arguments that can be held simultaneously in working memory and weighed against each other differs from subject to subject. Some people are able to keep up to seven arguments simultaneously in working memory, whereas others are more constrained and can use only four or five at a time. Irrespective of these variable constraints, the inevitable conclusion from these considerations is that the outcome of a decision process is the *only possible outcome* in this moment. Only if there are two equally probable future states of the network, which is extremely unlikely, would small random fluctuations of the system's activity become influential and determine which bifurcation the system will follow.

Matthieu: Fine, but once the decision becomes conscious—"I want to do this, I want to steal, I want to lie"—even if that decision has been built up unconsciously in the brain and you have been driven to it, a regulatory process also takes place and says, "Hey, do you really want to do this? This doesn't feel right." I may sense that I cannot resist my urges and yet still struggle against them, so that a regulatory process takes place and modifies or overrides the initial decision. Such regulatory processes exist and can be called on. They allow for emotional regulation. They can also be strengthened by trying repeatedly to exercise this regulation, reflecting on the negatives consequences of your impulses on others and yourself, being inspired by role models, and so on. Then a strong aspiration might come to your mind: "I really should not do this." So although you may not be "responsible" at a particular moment for wanting what you want and for the strong urges that arise, you have a certain responsibility to put this regulatory process into action, instead of avoiding or suppressing it. You are responsible for putting into action the steps to become what you want to be in one month, one year, or the rest of your life.

Wolf: The outcome of a conscious deliberation does of course have effects on future behavior. Following the experience that a certain decision had adverse consequences, a subsequent decision could be to change that particular habit. This decision and its effects, just like the aversive experiences following the first decision, will be inscribed in long-term memory and henceforth act as either subconscious motivations or conscious arguments, influencing subsequent decisions on how to act. Once past experience leads to a modification of priorities and behavioral strategies, the system strives to reach these new goals. Apparently the organization of our brains is such that feelings of uneasiness are generated if set goals are not pursued. It seems to be the same uneasiness that drives us to resolve conflicts. In this way, repeated experience with the outcome of decisions, positive or negative, can eventually induce a lasting change in the functional architecture of the brain and hence its behavioral dispositions. This in turn will impact the outcome of future decisions.

One should keep in mind, however, that both the initial decision and the inscription of the goal in memory as well as the bad feelings associated with the nonpursuit of the goal are the *result* of neuronal processes and not their cause. It is the neuronal process that evaluated the outcome of the first decision that caused the inscription of the new goal in memory. This new engram, this new memory trace, alters the state of the brain and influences future neuronal decision processes.

Eventually, the newly set goal, which initially may have had the status of a conscious rational argument, may change its status and become a habit that henceforth influences behavior without having to appear in consciousness. It can become one of the variables that acts at a subconscious level. One then refuses another glass of wine simply because it does not feel right to drink too much.

Matthieu: We can certainly improve our emotional regulation through learning. Play among children, even rough play, is known to be part of that. When they engage in rough play, for instance, children and young animals learn when to stop before really hurting the other young. One of the roles of the higher functions of the human brain that might not be found in other mammals—although I don't know much about it—is precisely to allow for a sophisticated form of emotional regulation, which again is a form of responsibility.

But I was not only speaking about improving our emotional regulation over an extended period of time. My point was that at any time, even when we feel a strong urge to do something, we have the capacity to evaluate the degree of desirability of the action and master our will and mental strength to refrain from engaging in this unwholesome action, even if the urge is strong.

Wolf: Yes, evidence indicates that more highly evolved brains possess hierarchically organized control systems that allow them to render responses that are increasingly dependent on internal variables and not solely on external stimuli. This increases the degree of freedom the brain has when responding to environmental conditions and also allows the brain to take the initiative rather than simply reacting to disturbances. Accordingly, the more actions and decisions are subject to these internal control systems, the more we are inclined to attribute responsibility to the agent. In children, some of these control systems are not yet fully developed, and therefore we hold them less responsible for their actions than adults. Along the same line of reasoning, adults are typically considered guiltier for decisions and actions that they have come to by conscious deliberations than for actions they have performed in an unreflective, thoughtless way or as a spontaneous response to a stimulus. One does not blame a child for acting out spontaneously, but if one realizes that the child has understood the rules of conduct and still acted against them, one tends to say that the child is responsible for what she did and deserves sanctions.

Matthieu: What you are saying is that when you sense that a behavior is unwholesome, you are aware that you could use your ability to regulate your emotional impulses and yet still you ignore both of these facts, your liability or responsibility seems to be greater.

Wolf: In this context, it is interesting that most people, as well as our legal systems, tend to associate responsibility mainly with consciously planned actions. One reason may be that the rules of conduct are made explicit in society by means of language that can only be decoded by conscious processing. Admittedly, much of early education is nonverbal, resembling the conditioning procedures used to shape the behavior of domesticated animals. However, once language understanding starts, most of the rules and imperatives of conduct are communicated through verbal instructions and rational arguments. One is supposed to be aware of them and be able to include them in one's deliberations. If one violates those rules, then one is held guilty because one is supposed to respect them during the decision-making process. Again, we distinguish between actions resulting directly from one's subconscious drives and those that we initiate intentionally after some conscious deliberations of the pro and cons.

Decision making occurs at two different levels, as we discussed briefly before. Most of the decisions that get us through daily life rely on subconscious processing and follow well-adapted heuristics. If these decision processes do not lead to immediate action, they may still influence subsequent behavior by manifesting themselves as what we call "gut feelings." One has no conscious recollection of the reasons that lead to these feelings, but one clearly experiences the reactions of one's autonomous nervous systems when the results of subconscious processes are in conflict with the outcome of conscious deliberations. In such situations, one tends to think, "I made the best decision I could according to all the rational arguments that I know, but it somehow still feels wrong." The other way around is also possible: "I did what felt right to me, but if I think about it, it was absolutely crazy and irrational." It is also clear that one feels good, satisfied, and to some extent "free" when the two decision systems converge to the same solution.

Matthieu: That coherence between reason and deep feelings could be enhanced and preserved through mind training.

Wolf: Kant said that if you can internalize the rules that are imposed on you, the external imperatives for ethical conduct, such that they become your *own* imperatives, then you are at peace. We would experience this more if it were possible, as you said, to enhance the congruence between unconscious and conscious processes through mind training.

Matthieu: There can be a serious problem when the social imperatives set from the outside are not truly ethical, in the sense of being attuned to the well-being of others, but rather are dogmatic and oppressive, as may happen under a totalitarian system or in some of the oppressive ancestral traditions that led to slavery, human sacrifices, domination of women, and so on. In such cases, it would be quite good and reasonable to feel at odds with the outer imperatives and not accept them blindly.

Wolf: This point is extremely important. What if subjective coherence is achieved by making the *bad* part of our behavioral dispositions concordant with distorted, unethical outer imperatives? History—especially Germany's recent past, modern terrorism, and many other crimes—provides a host of deplorable examples. When society reinforces the in-group–out-group dichotomy and declares the out-group members hostile, evil enemies, it calls on all the instincts that we have inherited that were initially meant to be useful for the defense of one's own kin. Once this manipulation of cognitive schemata has been accomplished, aggressive acts that would be considered highly unethical when performed against in-group members are perceived as moral duties when directed at out-group members. Hence, tribal warriors, crusaders, nation-defending soldiers, and, alas, terrorists and torturers often act in accordance with outer imperatives and inner drives, experience no conflict, and may even pass as heroes rather than murderers within their own community.

Matthieu: The philosopher Charles Taylor wrote, "Much contemporary moral philosophy... has focused on what is right to do rather than on what is good to be, on defining the content of obligation rather than the nature of good life; and it has no conceptual place left for a notion of the good as the object of our love and allegiance or as the privileged focus of attention and will."[3]

As Francisco Varela also wrote, a truly virtuous person "does not act out of ethics, but embodies it like any expert embodies his know-how; the wise man is ethical, or more explicitly, his actions arise from inclinations that his disposition produces in response to specific situations."[4] When you are confronted with a sudden situation that is evolving very fast and there is no time to ponder what to do, what you do spontaneously is the outer reflection of what you are internally at this moment in your life.

Spontaneous, "gut feeling" morals arise from our deepest positive qualities or flaws. In the case of positive qualities, they cannot be built exclusively on intellectual ideas but must be the expression of our having integrated in our mind streams benevolence, empathic concern, compassion, and wisdom. Such qualities can be trained like any other skill.

Wolf: If this is true, then it would imply that we can overcome by training some of the deep-rooted drives that we inherited from evolution, behavioral patterns that were well adapted to ensure survival and reproduction in a precultural world but that may no longer be appropriate. I hope you are right because it is absolutely essential for the future of humankind to render these negative drives less influential.

THE RESPONSIBILITY TO CHANGE

Matthieu: We may have no choice now about what we are; otherwise, everyone would certainly choose to be someone filled with admirable qualities rather than a criminal or a sex addict, an object of contempt for others. We may also have no choice about the way we behave in the spur of the moment. But we have the responsibility to change when change is desirable, and we are responsible, to some extent, for not having engaged in a process of transformation in the past.

When you recognize that you are not in control of your emotions and it is causing you suffering, it is your responsibility to escape from this trap.

Let's take the example of someone who has a strong compulsion to do something. As you say, when she is overwhelmed by her drives, she may not have much immediate choice or sufficient capacity to control her actions. As Oscar Wilde said, "I can resist everything, except temptation." But if that person knows that she has traits or tendencies that are detrimental to others' and her own well-being and has experienced her lack of control in the past, suffered from it, and caused others to suffer from it, she could take these experiences as opportunities to change. There are certainly times where the circumstances are such that the compulsive tendencies of that person are not expressed in such a strong way. Wouldn't that have been the time to do something about these traits by training the mind through using proper antidotes or seeking help from people who could offer means and methods to achieve that change? Doing something when it can be done and seeking the right kind of help is part of our global responsibility.

We might behave like robots in the spur of the moment, but we are not robots over the course of our life. Because everything is the result of causes and conditions, when all the causes for an event, whether in the brain or elsewhere, are gathered, that event has to happen. Yet over time we can generate new causes and conditions and influence this dynamic process. This is the virtue of mind training and of brain plasticity: Being exposed to new conditions induces brain changes, including in so-called unconscious processes.

One may disapprove of an action, but people are not intrinsically evil, although their actions may be evil. However people think and behave is the result of a dense web of causes and conditions that are changing naturally and can be further changed through specific interventions. Every person is more or less confused, more or less deluded or "sick" in his or her mind, and we should approach everyone as human beings who have gone through countless experiences under countless outer and inner influences.

Blame often proceeds from ignorance, contempt, and lack of compassion. A physician does not blame his patients, even if they have behaved in ways that have harmed their health; he tries to find ways to cure them, to skillfully help them change their habits. When someone harms others, he should be prevented from doing so with adequate, efficient, and measured means, but he should also be helped to change his harmful behavior.

Instead of engraving judgments about people in stone, we should view them—and ourselves as well—as flowing, dynamic streams that always have a genuine potential for change.

When Nelson Mandela was asked how it was possible that he made friends with his jailers during his 27 years of detention, he answered, "By bringing out their good qualities." When asked whether he thought that all people had some good deep within them, he answered, "There is no doubt whatsoever, provided you are able to arouse the inherent goodness."

Wolf: You are absolutely right. Human beings possess a wide spectrum of behavioral dispositions ranging from the extremes of cold-blooded murder and genocide to altruistic self-sacrifice. There are several ways to bias behavioral outcomes. One is to change the outer imperatives, to design socioeconomic systems that reward behavior that contributes to the stabilization of the system and sanction destructive actions. In a sense, this resembles an evolutionary process that eventually leads to the emergence of social interaction networks whose architectures favor adapted behavior. However, humans also go through a process of cultural evolution in which the transmitted traits are encoded not in genes but in codified moral conventions that get expressed in social conduct and habits. The important difference is that cultural evolution has an "intentional" component because we intentionally design the interaction architectures that impose the constraints for adaptation and selection.

Another strategy is to codify moral values and design education systems by which these values can be transformed into action-constraining imperatives that are internalized by individuals and transmitted from generation to generation. These are then complemented by normative systems that put further constraints on the range of tolerated behavior. If I understand correctly, there is yet another option, apparently favored by the Buddhist traditions, consisting of attempts to alter the behavioral disposition of individuals through mindfulness training. In all cases, the goal would be to eventually obtain as much agreement as possible between outer imperatives and inner dispositions, both forces being directed toward the good. The functional architecture of our brains can be modified by education, positive and negative experiences that act as rewards and punishments, insight, training, and practice. The incentives to change are provided by the reward systems in our brains.

Our brains seem to be endowed with systems that are capable of identifying brain states as consistent, coherent, congruent, and harmonious or conflicting and unresolved. We still don't know what the neuronal signatures of these states are, but we strive for the former and try to overcome the latter.

Matthieu: Those words—"coherent," "harmonious," and so on—refer to categories of experience not neuronal processes. This shows that to make sense of all this, you need to resort to your own subjective experience.

Wolf: Not necessarily. Subjective feelings of harmony must be associated with particular neuronal states, and it is possible that coherent neuronal activity patterns evoke positive feelings. As mentioned before, we still ignore the signatures of activation patterns that correspond to solutions and conflict-free internal states. Maybe such states are actually characterized by a high degree of coherence. Because we are social beings deeply embedded in our cultural networks and permanently exposed to the judgments of others, the criteria of consistent internal states are derived from the heritage of biological evolution as well as the imperatives that emerged from cultural evolution, the moral standards that we are supposed to meet and that get installed in the architecture of our brains by education, observation of social interactions, and life-long exposure to social rewards and sanctions. Once internalized, these experiences assume the status of internal goals, as self-imposed imperatives that need to be reached to arrive at consistent, nonconflicting internal states—just as we strive for the satisfaction of biologically motivated goals such as satiety and reproduction. It all depends on the nature of the values and norms that our cultures impose on us and that, once installed in the functional architecture of our brains, act as goals and imperatives.

Values and norms installed early in life, before the development of episodic memory, remain implicit and deeply anchored in our subconscious, and therefore they are experienced as integral parts of our personality. They are experienced as our own goals, drives, convictions, and moral judgments. Set points imposed on us during later life are more often explicit; we are conscious of their origin and therefore experience them as imposed social constraints that need not necessarily agree with our internal convictions. Nevertheless, we wish to comply with them to reduce conflicting states in our brains and strive for a coherent state.

Matthieu: Surrounding cultures are not only imposed on us. We also fashion our culture through our thoughts, personal transformation, and intelligence. Individuals and cultures are like two blades that sharpen each other. Because both contemplative science and neuroscience have shown us that it is possible to train our mind and gradually modify our traits, the accumulation of individual changes can also shape a new culture.

Wolf: Obviously we create cultures, and they in turn act on us. I also tend to agree that it is possible to override some of the inherited and early imprinted traits not only through cognitive control—a strategy that we all apply in adulthood—but also through procedural learning by simply developing new and better adapted habits. I see an analogy here with the acquisition of new motor skills, which after sufficient training can be executed without cognitive or conscious control.

Matthieu: This is quite similar to what contemplatives say. In the beginning, any practice is contrived and unnatural. With familiarization, we begin to do it well and with ease, and, finally, it becomes fully part of us.

Wolf: I could therefore imagine that it is possible to also change certain traits with practice. Resisting a particular temptation or engaging in an altruistic act, both of which would be rewarded by social recognition, initially requires cognitive control and the investment of attentional resources. It is conceivable that practicing these novel attitudes over and over again could eventually anchor them in brain structures that execute their functions without requiring cognitive control. In this case, the novel behavior would become more and more a new trait of character. We observe such changes in attitude in biographies, and it would be interesting to examine experimentally the extent to which these changes can be induced, even in adults, by self-paced practice and training, or whether they are more likely the result of singularities, such as trauma or enlightening experiences. The evidence that wisdom grows with age testifies that accumulating experience can change character traits, so there is hope that training compassion and generosity can be effective.

Matthieu: If a certain number of individuals can thus undergo personal change, this will naturally induce gradual changes in the surrounding culture—

Wolf: —which will have a reciprocal impact on individuals. This reciprocity could, at least in principle, initiate self-reinforcing progress toward better societies, just as, in the opposite direction, aggression and revenge initiate vicious circles that have deleterious effects on societies. We should exploit both options for change: working on the individual and designing interaction architectures for societies that provide incentives for peaceful behavior.

FREE WILL AND THE RANGE OF CHOICES

Wolf: But let us return to the question of the relation between the concepts of free will and guilt. If somebody got in trouble with the law because he acted in a way that is considered inappropriate given the societal context, then one of the first questions to be answered is whether the person had full control of his cognitive abilities and was capable of comprehending the nature of the act and considering its consequences. One examines whether the ability to reason and weigh all possible arguments was unimpaired. One also examines whether the delinquent was intoxicated, extremely stressed, tired, or under pressure to hasten the decision or whether cognitive abilities were hampered by a psychiatric or neurological disease. If none of these attenuating factors can be identified, then the delinquent is considered fully responsible because nothing was found that could have impaired his capacity to consciously consider all the arguments needed to evaluate the appropriateness of his decision.

Yet the range of options available at any particular moment in time varies widely. The range of options can be large when there is no external pressure, no compulsive internal drive, when consciousness is fully awake, and there is ample time for deliberations and examining all possible outcomes. However, even in this optimal case, the taken decision is the only possible one; if another decision had been taken, then it would again have been the only one that the brain could come up with at that moment under these circumstances. Schopenhauer clearly recognized that we decide according to our wish and will, but we cannot want otherwise than what we actually wanted in the moment of deciding.

Matthieu: Assume that we don't make use of our ability to consciously exert control over our impulses. Don't you risk ending up with nothing more than a tautology: "At any moment, only what *is* can *be*"? Surely you cannot claim that at any moment, what is *could not have been otherwise*. It is quite true that there is no point in denying what already is or demanding that what already is should be different. Yet we could possibly have prevented its happening, and we can certainly prevent it from happening again. This can be achieved, for instance, by acquiring new knowledge about what is desirable or undesirable to do and by training our minds.

Wolf: If initial conditions are changed (e.g., by creating new memories), then the outcome of a decision will be different. If I recall that decision A got me into trouble, then I will try to avoid it if I am exposed to a similar condition again. Yet here comes the problem that I wanted to address: Our legal system posits that people are in principle free to decide. If they do not decide in the right way, they are considered guilty, and the depth of the guilt depends on the options available at the moment of the decision. Put another way, the amount of attributed guilt is dependent on the amount of attributed freedom. However, "freedom" in this context only extends as far as the range of options available at the moment of decision making. If this range of options is small because of identifiable inner or outer constraints, then one could argue that the person did not have much of a choice and any other person in this situation would have decided and acted similarly. Thus, I suspect that what judges do in practice is not to evaluate the philosophical question as to whether our will and volition are free and unconstrained but to simply examine the extent to which a person's decision or action deviates from the norm. They explore as far as is possible the inner and outer constraints at the moment of the decision and then infer the magnitude of the deviance by assuming what so-called normal people would have done under comparable conditions.

In other words, what judges do is to determine how much the functional architecture of the delinquent's brain differs from that of average citizens. If the conclusion is that, given the conditions, the average citizen would likely have acted in the same way, then attenuating causes will be granted and the sentence will be mild.

Interestingly, this reasoning is based on the implicit assumption that subjects decide and act in a rather predictable way as a function of conditions, which tacitly implies that the decision process is influenced by causes. This finding suggests that our legal systems are in fact not based on the fiction of an unbound free will. It is just a convenient assumption that agrees with our subjective experience that we are free at all times to decide otherwise. Thus, even if we abandon this fiction, even if the neurobiological evidence against the existence of an unconstrained free will eventually receives widespread recognition, this will not jeopardize the legitimation of our legal systems to enforce compliance with set norms. We shall continue to attribute responsibility in the sense of authorship to individuals and sanction violations of the norms. The notion that a person's decision is the result of neuronal processes that obey the law of causality and that there was no possible alternative to a decision at the time it was taken does not imply that the person is not responsible for his or her action. Who else could be to blame?

It is clear that there is no strict correlation among the attributions of free will, responsibility, subjective guilt, and severity of punishment. If one runs a red light because one had an attentional blink but nothing happens, then one is penalized by a few points on one's driver's license and gets away with a monetary fine. However, if the same attentional blink, the same subjective error, becomes the cause of an accident and people are killed or severely injured, then the sentence will be much more severe. Thus, our legal system also considers the severity of the outcome of an act, not just the act itself. This satisfies our conception of justice and fairness: If somebody has caused much suffering, then we consider that there must be some retaliation to reestablish equilibrium.

Matthieu: This is why the risk is that justice becomes a system to legalize revenge rather than to impartially weigh the nature of motivations, intentions, and true responsibility. As in the example that you gave, the severity of the outcome is not necessarily related to our intentions and is often unpredictable and beyond our control. At the same time, a sense of responsibility comes along with simply being the agent, whether voluntary or not. If at a friend's house I accidentally knock down a beautiful porcelain vase and it breaks on the floor, I feel sorry, I apologize, and I will be eager to replace the vase. I consider myself responsible even if I had no intention at all of breaking the vase, and it was simply caused by my lack of vigilance or some unforeseen circumstances—I suddenly moved to the side not to step on the cat sleeping on the floor, which I had not seen early enough, and in my awkward movement knocked over the vase or what have you. However, I also could have been more mindful and attentive while moving around in an unknown environment. There are some clumsy people who knock things over a lot, but once they know this about themselves, their responsibility is to be extra careful. If I am not involved at all and see no connection between the incident and me, for instance, if the shelf that holds the vase collapses on its own in the same room as me, then I don't feel any kind of responsibility.

ATTENUATING CIRCUMSTANCES

Wolf: I agree. Even if our will is not as free as our intuition suggests, then we are of course responsible for what we do because we are the agent; decisions are our decisions, and acts are our acts. We are the authors, and just as we want our merits to be attributed to us and rewarded, we also have to accept the sanctions for our misconduct. This assignment of responsibility is by no means invalidated if we abandon the fiction of free will.

But let me briefly return to the matter of granting attenuating conditions. Let us imagine that a capital crime has been committed and there were no attenuating circumstances—the murderer had enough time to deliberate and was not under any special internal affective or external pressure, the spectrum of his options was wide, and his only identifiable motive was some minor argument with his victim. Accordingly, he will receive close to the maximum sentence of decades of jail. A few months later, the delinquent has a seizure. His brain is scanned, and doctors discover a tumor in the prefrontal lobe. Immediately, the case is reconsidered. The tumor had probably destroyed centers in the orbitofrontal cortex where moral values are stored and cognitive control is orchestrated to suppress actions judged as inappropriate. In young children, these structures are still immature, and therefore children are less able to withhold actions. The delinquent, now considered a patient, lacked the neuronal machinery to control his actions through the function of stored moral values. He will be transferred from jail to a clinic. This change in perspective was possible because of a clinical symptom and the availability of a powerful diagnostic tool.

However, if there is no evidence of disease in terms of clinical diagnoses, a neurobiologist would still have to confirm some abnormality in the delinquent's brain because he did what no average citizen in our society would have done. Many possibilities can be considered, none of which can be diagnosed with presently available tools. The circuits within or between the centers required for the storage of moral values and response inhibition could have developed abnormally for either genetic or epigenetic reasons. The same could have been the case for the learning mechanisms that support acquisition of social norms. Deficiencies in education could have occurred so that the respective engrams, the neuronal changes underlying memories and required to store social imperatives, were not sufficiently consolidated. And so on.

Matthieu: Although the diagnosis is not as dramatic as a tumor, the result is the same: Something is wrong with his brain.

Wolf: The result is the same, but the causes and hence also the consequences for the delinquent differ. If the neuronal causes are identifiable, then the delinquent becomes a patient; if they cannot be specified with the available tools, the delinquent goes to jail.

Matthieu: In both cases, we should consider the delinquent as being a sick or at least dysfunctional person and, while preventing him from going around stealing things or harming others, what we should really do is adopt the perspective of a physician and treat and help him through appropriate means. This accords exactly with the Buddhist view, which says that we are all sick because of ignorance, greed, hatred, craving, and other mental toxins and therefore need to follow the advice of a skillful physician—a qualified teacher—to undertake a treatment of inner transformation so that we may be cured of those mental toxins. You don't retaliate against people for what they do under the sway of mental illness. They need help from outside, from the wise and experienced persons who have insights on how all this happens and can show them the methods to change. From their side, they need the intelligence to recognize the need to change and the determination to use appropriate methods to gradually bring about the transformation.

Wolf: For this to be possible, novel values and norms, novel internal goals, need to be installed in the defector's brain. Such new set points can emerge naturally if the outcome of a decision is aversive for the subject or gives rise to regret. However, there will also be cases where new set points need to be installed from outside because his or her "sick" brain cannot generate these goals. The brain of somebody who kills out of an impulse of wanting money *must* differ from that of somebody who does not. It appears then as if the court judges the functional architecture of delinquents' brains. Those whose brains deviate far from the normal distribution are punished most severely, whereas those with brains that produce well-adapted behavior that conforms with social norms most of the time are treated more mildly because persons with such brains will break the law only under conditions considered attenuating.

This viewpoint by no means questions the responsibility of the actor. I cannot emphasize this point enough! Who else than the actor is responsible for his or her actions? What needs to change is *our attitude toward people who fail.* Their actions must be disciplined, even if we recognize that they could not have acted otherwise given the conditions. This point caused much confusion when I first exposed these arguments. The false conclusion was, "If the delinquent could not have acted otherwise, he is neither responsible nor guilty and therefore cannot be punished. In consequence, there will be anarchy: everybody could simply do anything he or she wants." This argument is of course nonsense. In some sense, we need the same process we already take with children: They lack essential control mechanisms, and so we are lenient but still punish and reward them to change their behavior to install novel set points and goals in their brains that they can then pursue.

Matthieu: Therefore, instead of retaliation, we should lay the emphasis on education, rehabilitation, training, and personal transformation while also taking appropriate measures to prevent wrongdoers from harming others.

Wolf: I have often been accused of attacking human dignity when arguing against the fiction of free will. Frankly, I cannot see why the attempt to communicate neurobiological insights should challenge the dignity of human beings. It questions neither agency nor responsibility for one's acts; it rather changes our perspective and provides us with a better understanding of human behavior and a more informed and sincere motivation for taking proper care of people and sanctioning aberrant behavior. In essence, it should lead to more humane, insight-driven treatment of the unfortunate human beings who have to live with a brain that produces behavior that is in conflict with social norms.

Matthieu: That's an interesting and thorough argument that has quite a few similarities with the Buddhist perspective.

Wolf: This is good to know.

LOOKING WITH THE EYES OF A DOCTOR

Matthieu: As I began to mention, in Buddhism we consider ourselves sick people who err in the cycle of conditioned existence, what is called *samsara*, or the world of mental confusion and ignorance.

Wolf: This is like the idea of inherited sin in Christianity, right?

Matthieu: No, it is quite different. It is not a fundamental trait of human nature. It is rather due to having lost sight of our fundamental nature. When you are sick, you say, "I have the flu." You don't say, "I am the flu." So the sicknesses that cause suffering are that of hate, craving, and other mental toxins. These sicknesses are not intrinsic but result from ever-changing causes and conditions. Sickness is not the normal baseline of a living being but an anomaly that reduces our chance to survive. According to Buddhism, the normal healthy state, the basic human nature, which is also the fundamental nature of mind, when it is not obscured by mental clouds, is more like a nugget of gold that remains pure even when submerged in thick mud. Buddhism leans more toward the view of original goodness than that of original sin. This does not mean that hatred and obsession are not "natural" and are not part of the repertoire of the human mind. We know that well enough!

Rather, it means these afflictive states of mind result from mental fabrications that obscure our understanding of the basic nature of mind, pure consciousness, as the ore conceals the gold it contains. We need to differentiate the fundamental nature of our mind from the various afflictive mental states that lead to all kinds of suffering. Accordingly, no one in this world is fundamentally evil but rather sick because of the effects of mental poisons. A person is not fundamentally defined by his sickness.

Wolf: But if you say the person has a criminal personality, that doesn't work.

Matthieu: Why?

Wolf: Because the personality *is* the person. Our personality, our character, and all the traits that are the object of social judgments are determined by our brains. Therefore, the distinction between an organic disorder and a mental disorder is blurred. A tumor in the liver is a somatic disease that leaves your personality untouched. A dysfunctional brain distorts the personality. In both cases, the person is sick. Unless we take a dualistic stance, we cannot distinguish between an uncontaminated, pure mind—your gold nugget—and impure, faulty neuronal machinery. It is possible to say that a person is "intrinsically" bad without denying, however, the potential for some change.

Matthieu: Not really. Even for someone who has a genetic predisposition for cancer, you would still say that a healthy condition is her normal state, whereas the disease is a pathological state due to various causes and conditions.

It is the same with your brain states and with the nature of mind. The fact that the mind can be invaded by mental toxins, which are mental constructs, does not mean that it could not have been and cannot be otherwise. If you put cyanide in a glass of water, the mixture is a deadly poison, but the water has not been corrupted or made poisonous. You can purify the water, distill it, or neutralize the cyanide. The fundamental constitution of the water has not changed. Likewise, the basic nature of the mind can be obscured by afflictive mental states, but pure awareness can always be recognized beneath the screen of deluded thoughts.

Buddhism will say that even a person who has become the slave of hatred does not become fundamentally identified with hatred.

Wolf: But who is producing hatred and anger if not the brain that produces *all* the behavior that defines a person?

Matthieu: Hatred is produced by mental fabrications, chain reactions of deluded thoughts, and ignorance. One seemingly innocuous thought of irritation or resentment springs up in the mind, like a small spark, and gives birth to second and third thoughts, and soon the mind is filled with the fires of anger.

Wolf: Does this not imply a separation between the person, who is immaculate, and the emotions, in this case, the anger, that enslave the person? But where does the anger come from? If it is produced in the person's brain, then is it not a constitutive property of the person?

Matthieu: To get the flu is part of life and part of our physiology. It occurs within our body and affects us strongly. But it is not intrinsic to us. The flu is a transient state. Even if one considers things from an evolutionary perspective, individuals and species are selected for traits that favor their survival. All ailments are anomalies that are unfavorable to survival.

Likewise, the nature of the mind is pure cognition or basic awareness. It can be populated by all kinds of content, all of which are impermanent and changeable. The potential for change always exists.

Wolf: If a person has bursts of anger that eventually get cured through therapy and disappear, would you say that his or her personality has changed?

Matthieu: The person's traits will have changed for sure, but not the basic nature of awareness. We should not equate the temporary state of the person's mind or brain that leads to uncontrollable anger with the whole mind stream of that person over its long history. No one is intrinsically bad because mental constructs are impermanent. A river can be polluted, but the river can also be purified. In the case of a mind stream, it may take time to change its contents, and this process might be more or less difficult to achieve, but the potential for change always exists.

Of course, you have to do something about it. You cannot just snap your finger and say, "May this water be purified!" or "If it does not become purified right now, I'm going to throw it away." Likewise, when you sentence someone to death, you throw away the water of his life without giving it a chance to change. The person may have no choice to be what he is now because he is the result of many past events, but he has a choice of who he will be in the future and, consequently, a responsibility to undertake a process of change.

Wolf: So you would favor the view that somebody did what he did because his brain didn't allow him to do otherwise at that moment?

Matthieu: In the spur of the moment, someone with an untrained mind might be truly unable to control her anger. But, realizing this, the best thing she can do for herself and others is to undertake a process of mind training.

Wolf: What if the person lacks the cognitive ability to perceive the aberrant conditions or the force to induce a change?

Matthieu: Change does not occur easily and all at once. It is a matter of beginning a process that will bring gradual change, with patience and perseverance. Of course, for that to happen, there must be at least some eagerness on the part of the dysfunctional person. Here, too, one may help that person become aware that he is not "fundamentally bad" but some unfortunate aspects of his mind and behavior bring all kinds of suffering on others and himself, and he would be much better off if he accepted the idea of undertaking a process of change.

TRUE REHABILITATION

Wolf: So punishment would be required as an incentive for change, as an instrument for education, but it should not satisfy feelings of revenge?

Matthieu: Yes. Revenge is fundamentally wrong. It proceeds from animosity or, in its extreme form, from hatred. It is willingly becoming infected by the same sickness as the person you're calling sick. As the president of the Italian Coalition to Abolish the Death Penalty, Arianna

Ballotta, said, "As a society, we cannot kill to show that killing is wrong." If hate answers hate, then hate will never cease.

Wolf: Is punishment an attempt to reestablish justice, reinstall equilibrium, and deter future delinquency?

Matthieu: Punishing someone is not the same as taking revenge as long as it is done with a compassionate, educational purpose, as parents do with their children.

Wolf: Is it a means of self-protection?

Matthieu: This is a different aspect of the question. While reeducating wrongdoers, one needs to protect society from their wrongdoings and also protect the perpetuator against his own sickness. Jails should really be rehabilitation centers, catalysts for change. Unfortunately, they rarely fulfill that function and are most often not only places of punishment but nests of violence and antisocial behavior, constantly reinforcing the inmates' deleterious propensities and creating a culture of abuse. At best they are a means to isolate dangerous people from society, which is better than revenge and punishment, but they mostly fail to address the possibility for change. As long as someone can't be cured, it is right to prevent him from harming others, but this should be done without any hatred or feeling of vengeance. The examples of Scandinavian prisons, for instance, where the rehabilitation of offenders is the priority, has shown that once offenders are freed back into society, the incidence of recidivism is much lower than in countries where there is a stronger culture of punishment.

I heard a devastating report from the BBC about juveniles who, until recently, were judged as adults in several states in the United States, which goes against international laws. In Denver, Colorado, there are a number of such kids. One 16-year-old who helped a friend cover up a murder was tried as an adult for a felony crime and sentenced to life imprisonment without parole. Although the law has changed now, following a decision from the Supreme Court in May 2010, it is not retroactive, and this adolescent will spend his whole life locked up in jail for helping to cover up someone else's crime. In the United States, because of voters' pressure to keep streets safe, local district attorneys often decide on the fly that a youngster will be tried as an adult and will not even get the chance to go to court before a grand jury. In the case of this 16-year-old, he was sentenced to life imprisonment in a matter of moments. No matter what, he will never get out of jail in his lifetime. The BBC interviewed this young person, and he said that he had succumbed to the pressure of his friends to help cover the murder, in which he did not even remotely participate. He said in the interview, "It just happened. I was there, he was my friend, I didn't know what to do. We panicked. I helped him to cover it up and we ran away."

Wolf: All this, even though we know that the brain develops at least until the age of 20, maybe even 25, and that teenagers are notoriously challenged by emotional surges.

Matthieu: This kind of punishment is a tragic mockery of justice and totally fails to recognize the vast possibility of change in someone, especially in a young person, who has not even had the chance to fully build up his or her personality and emotional control. It also prevents the perpetrator from engaging in any activities that would attempt to repair the harm he has done, which is one of the best ways to bring back harmony to society.

This kind of populist behavior by the local attorney is clearly not inspired by a fair sense of justice but by the desire to satisfy the public's visceral thirst for swift and spectacular punishment. In fact, it is revenge. Of course this does not at all mean to ignore the fate of the victim and anguish of his family. But the judge's actions disregard all we know about people's potential for transformation, which is attested by neuroscience and supported by the Buddhist view.

If we want to achieve a more compassionate society, then we need to provide the possibility for all—perpetrators, victims, and judges alike—to change their attitude, reactions, and the way they treat others. This would be a much better way to keep our streets safe. Locking 16-year-olds up in jail for life won't do it. There are indeed repeated offenders, from whom we must protect the population, but there are also people, including these youngsters, who acted under extreme pressure and regretted immediately and deeply the tragic blunder they made.

Wolf: And those who would never have done it again. Their brains are probably not far from the norm and receptive to educational measures.

Matthieu: We know in particular that the death penalty has not proven to be an effective deterrent. Its elimination in Europe was not followed by a rise in crime, and its reestablishment in some American states was arguably not followed by a drop. Because imprisonment is enough to prevent a murderer from committing further crimes, the death penalty is therefore nothing but legalized revenge. Compassion is not a reward for good behavior, and its absence is not a punishment for bad behavior. Compassion aims to remove all kinds of suffering, whatever they are. One can make moral judgments, but compassion is in a different category entirely. Compassion for the victims is to assist them in all possible ways. Compassion for the perpetrator is to help him to get rid of the hatred and other mental dysfunctions that led him to act in harmful ways. This in no way minimizes the harmfulness of his deeds, but rather recognizes the possibility for change, repair, and forgiveness.

Wolf: I often end my talks on free will with these same conclusions. These are the consequences of abandoning the fiction of free will, and I consider them humane. Still, I continue to be attacked because people feel that human dignity is at stake, I am betraying the values of our culture, and it is an anarchistic position. On the contrary, if we adopt the view that there is a neuronal reason for deviant behavior, whatever its causes, genetic or epigenetic, then we are bound to exclude any aspect of revenge, retaliation, or compensation from our legal practices.

Matthieu: But I wonder whether simply adopting the view that there is a neuronal reason for deviant behavior can give rise to genuine compassion. One might simply adopt a cold-blooded, "objective" stance, which might lead to not seeking revenge but might also not engender benevolence or compassion.

Wolf: I disagree! Once we come to consider delinquents as patients, as victims of genetic, epigenetic, or disease-related disturbances of their brains, it becomes easier to treat them with benevolence and compassion.

Matthieu: If one sees it that way, it is fine indeed, because it will allow one to move beyond one's instinctive reactions when faced with the behavior of malevolent people and act like a physician, not an avenger. If a patient suffering from mental disturbances strikes the doctor examining him, the latter won't hit back but will seek out the best ways to cure him of his illness.

HORRENDOUS DEVIATIONS

Wolf: Recognizing that all behavior, including ethical judgments, has a neuronal substrate could also enhance the humbleness and gratitude of all those many healthy people who are lucky enough to have stable dispositions that are unlikely to produce deviant behavior. However, we should not forget that even seemingly healthy brains can be radically reprogrammed. Think about the many fathers from peaceful German families who turned into supervisors, cold-blooded murderers, in the Nazi camps. It took only a few years of a campaign of ideology, propaganda, and brain washing to accomplish this goal. This scenario demonstrates the danger of the plasticity of human brains, even those that probably did not differ much from the average. These unimaginably monstrous crimes deserve the most severe punishment, and I must admit that it is difficult, if not impossible, for me to consider these murderers as patients having abnormal brains.

Matthieu: You are a good person, capable of compassion and feeling indignation in the face of barbarity and injustice. So you react as a human being, and you realize that the strictly neurobiological explanation of human behavior and relationships does not tell us enough. These Nazis were normal human beings, with normal brains, but they were filled with ideological hatred and encouraged to work objectively, coldly, with no compassion whatsoever, for Aryan domination. They were suffused with "bio-eugenistic" ideology, which allowed them to maintain a sense of good conscience when they killed people. What they were missing was not biological knowledge but compassion! What they deserve themselves is not hatred but compassion as well—not a weak, permissive compassion, but a courageous one determined to counteract the causes of suffering, whatever they might be. To be deeply deluded is a mark of fundamental ignorance, extreme distortion of reality, and a lack of compassion and understanding of the law of cause and effect. To consider hate as acceptable or even promote it as a virtue is the archetype of mental delusion.

This does not necessarily mean that these people had abnormal brains, but they learned, sometimes quickly, to accept aberration as normality and become indifferent to the most horrendous cruelty. Many factors can lead to such extremes: exploiting people's fears and transforming these fears into hatred through calculated propaganda; counting on people's tendency to conform their behavior to that of the majority around them, even when that behavior becomes inhumane; blunting one's own empathy; demonizing others, thus eliminating all concern and respect for their well-being; or treating others like animals, ascribing no value to their lives. (It is following a similar process that animals are further devalued as mere objects of consumption without intrinsic value, other than commercial—and killed by the billion every year.)

Wolf: The Nazis were normal citizens before they converted. I fiercely oppose the interpretation that they just casually shifted into an abnormal state and therefore attenuating conditions were at work here, as I argued before. It seems that I am contradicting myself, not granting these monsters the same empathy that I allowed for other criminals. Why? Perhaps exactly because they were apparently quite normal people and obtained the motives and justifications for their atrocities from ideology and propaganda, to which they succumbed. Because they had proven before that they were capable of leading a decent and responsible life. Or were they in fact abnormal all along but managed to disguise their true nature until conditions allowed them to act out? Or could it be that what we consider as the norm is in fact a painstakingly maintained equilibrium between the potential evil within us and the constraining forces of moral imperatives imposed on us? If the default state is the potential for both good and evil, then is it not to be expected that the evil surfaces as soon as the embedding social conditions reward the evil and disparage the moral?

The attempt to cope with this human catastrophe in our own recent history also sheds a different light on the motives for punishment. Many of the people who were directly or indirectly involved in the realization of these atrocities managed to escape and led a normal life. In a forensic evaluation, they would have passed as no longer dangerous. Probably no attempt was made to reeducate these persons, and yet it is unthinkable that they would not have been pursued and sentenced for their acts. Revenge is an inappropriate term here because any possible sentencing would be dwarfed by the monstrous dimensions of their crimes. All we can demand is repentance and remorse. For now, I would like to conclude that understanding the processes that lead to criminal behavior is no reason to tolerate it.

Matthieu: Exactly. It's no longer a matter of objective explanation but of engaging with a human life. We must also understand that even if we believe that human beings' default mode is benevolence, deviation toward malevolence can happen easily and gain strength quickly. It is like walking on a mountain path. As long as you walk mindfully on the path, you are okay. But if you misstep over the edge, you may tumble down the slope before you even know what's happening.

We should neither be complacent toward crime nor decide that the criminal is fundamentally evil forever. We also need to understand that a person's behavior results from many complex, interdependent causes and conditions.

Wolf: Such a differentiated view lets the perpetrator sometimes appear as a victim of conditions. Are you saying we should forgive them?

BREAKING THE CYCLE OF HATE

Matthieu: Forgiveness is not saying that harmful actions are not bad and that a person will not have to face the consequences of his actions. Forgiveness is breaking the cycle of hate. It does not help to be caught in the same kind of hate that we want to punish. From the Buddhist point of view, there is no question of escaping the consequences of one's actions. The notion of karma is nothing but the application of the general laws of cause and effect to the consequences of people's motivations and actions. All actions have short- or long-term consequences. If you forgive someone and forgo retaliation, the person will still face the consequences of her actions.

Wolf: But how can you protect yourself against these genetically inherited emotional reflexes? Retaliation and revenge are deeply rooted human emotions and in premodern societies may have even had important functions for the stabilization of in-group solidarity. If somebody were to deliberately rape and kill my daughter, I am sure I would have a very, very hard time controlling my emotional outbursts and my desire for revenge and sanctions. These emotions give rise to hate and the desire for retaliation; because they are so fundamental, they are barely repressible by cognitive control and education. The critical question is whether mental practices can reach these deeper layers and change our emotional predispositions so that the desire for revenge does not manifest to begin with. If so, mental practice could indeed be more effective than our classical educational strategies because these seem to rely more on the imprinting of rules of conduct and the strengthening of the cognitive control mechanisms required to inhibit emotional reflexes rather than on changing the emotional basis.

Matthieu: After the bomb attack that claimed hundreds of victims in Oklahoma City in 1995, the father of a three-year-old girl who was killed in the bombing was asked ifwhether he wanted to see the perpetrator, Timothy McVeigh, executed. He answered simply, "Another death isn't going to ease my sorrow." Such an attitude has nothing to do with weakness, cowardice, or any kind of compromise. It is possible to be acutely sensitive to intolerable situations and the need to redress them without being driven by hatred. A dangerous offender can be neutralized by any means without losing sight of the fact that he is a victim of his impulses. In contrast to the father's attitude, the American radio station VOA News described the feelings of the crowd just as Timothy McVeigh was about to be sentenced: "People were waiting outside the building, silently holding hands. When the verdict came, they applauded and cheered. One said: 'I've been waiting a whole year for this moment!'" In the United States, a victim's family members are allowed to sit in on the execution of the killer. They often claim to be comforted by watching the murderer die. Some even assert that the condemned man's death was not enough and that they would have liked to have seen him suffer as badly as he made his victim suffer.

During World War II, Eric Lomax, a British officer, was captured when the Japanese conquered Singapore; he was a prisoner of war for three years and participated in the construction of the railway bridge over the Khwae Yai River (River Kwai) in Thailand.⁵ When guards discovered Lomax had drawn a detailed map of the railroad the prisoners were being forced to build, he endured intense questioning and torture, including water boarding. After his release, Lomax's mental anguish, hatred of the Japanese, and desire for revenge remained unabated for almost 50 years.

After undergoing psychotherapy in a center for victims of torture, he did some research and discovered that an interpreter who had been involved in the questioning, Nagase Takashi, someone whom Lomax particularly hated, had spent his life trying to make amends for his actions during the war by speaking out against militarism and engaging in humanitarian activities. Lomax was skeptical at first, until he found another article and a small book in which Takashi explained how he had devoted much of his life to atoning for the treatment that the Japanese army had inflicted on prisoners of war. He described the horrific scenes of torture inflicted on Lomax and how disgusted he was with himself for having attended these sessions. He spoke of having horrible nightmares, flashbacks, and painful trauma, just as Lomax had for decades.

Lomax's wife wrote a letter to Takashi saying that she hoped a meeting could take place between the two men to heal their wounds. Takashi quickly replied saying that he wanted very much to see Lomax. A year later, Lomax and his wife flew to Thailand to meet Takashi and his wife. From the first moments of their encounter, Takashi, his face awash in tears, kept on repeating, "I regret it all so much." Suddenly, paradoxically, Lomax was comforting his former interrogator, whose pain seemed more intense than Lomax's. Eventually, the two men came to laugh, recalling memories of the past, and found out that they now enjoyed each other's company. During the few days they spent together, at no time did Lomax feel a single flash of the anger he had long nurtured against Takashi.

A year later, Lomax and his wife went to Japan at Takashi's invitation. He asked Takashi to meet him alone and assured him of his total forgiveness. Takashi was overcome with emotion. Lomax wrote, "I felt that I had accomplished more than I could ever have dreamed of. Meeting Nagase has turned him from a hated enemy, with whom friendship would have been unthinkable, into a blood brother."[6]

In the face of such dramatic cases, I feel cause to believe that mental training can change your worldview and personal reactions. The only real enemy is hatred itself, not the person who falls prey to it.

IS THERE A SELF THAT BEARS THE RESPONSIBILITY?

Wolf: What you say makes me suspect that you assume a self that is immaculate, knows what it should do, has proper goals, and is not contaminated by drives and emotional dispositions inherited from biological evolution and cultural embedding. This proper self, if left to act on its own, would strive only for good, but unfortunately it cannot do what it would like to because it is constrained by evil forces, the negative emotions of anger, hatred, jealousy, and greed, the drive to possess, the drive to dominate, and so on. In assuming this, you dissociate the self from the rest of the person. To me this appears problematic because both the conscious self and behavioral dispositions emerge from the same brain. Consider, for example, the practice of blood revenge to reestablish the honor and pride of a family. This act certainly comes from a cultural norm rather than a genetically inherited trait.

Matthieu: Let me clarify. As you know, from the Buddhist perspective, the self is nothing but a mental construct that we use to name our mind stream. There is no such thing as a separate, autonomous, unitary entity that we could pinpoint as being the "self." So it is not the self that is immaculate; it is the fundamental nature of our consciousness, our basic, primary faculty of knowing, which is not modified by its content. If we are able to refer back to this pure mindfulness, then we have a way to deal with afflictive emotions.

Wolf: Is training for pure awareness really enough to get rid of all the deeply rooted traits that you equate with imperfection?

Matthieu: You train to become increasingly aware of the content of your mind so that you can rest in this awareness and continually recognize it without being carried away by your mental constructs and powerful emotions and without actively trying to suppress them.

Wolf: Well said. I think this point is extremely important because it clarifies the differences between our concepts of the autonomous self.

Matthieu: We need to experience this state of awareness and perceive awareness as being always present behind the screen of thoughts. We do have moments of peace, when we are spared for a while from the constant mental chatter that usually keeps our minds busy, when you sit quietly by the side of a mountain, for instance, or when you are exhausted after intense physical exercise. For a while, you may experience a quiet state of mind with few concepts or inner conflicts. This experience might give you a glimpse of what clear awareness, unencumbered by thoughts, might be. To recognize this basic component of awareness might give you confidence that change can indeed take place.

Wolf: Is this concept of the a priori purity of a consciousness that is segregated from the rest of a person's traits reflected in the legal systems of countries in which Buddhism prevails, such as Tibet, Bhutan, or Burma?

Matthieu: I was interested to hear from Bhutan's chief justice that the new Bhutanese constitution attempted to incorporate some basic Buddhist ethical concepts, in a secular way, to balance individual rights with duties and responsibilities toward society.

In ancient Tibet, as the Dalai Lama often emphasizes, things were far from being perfect, and there were some cruel punishments and practices. But these were extremely limited in comparison with the wholesale massacres that took place over the centuries in neighboring countries such as China and Mongolia. In Tibet, I have witnessed an old custom firsthand: When a feud occurs between two people, families, or tribes and resentment is flying high, the case is often brought to a lama by one of the parties. The lama then calls in both parties, talks with them, and eventually makes them take a pledge to stop retaliating against each other. To seal that pledge, he puts on both parties' heads a golden statue of the Buddha and has each of them take a vow of nonharming. I have personally seen people who a few moments earlier were just about to kill each other become calm again and sit together to sip a cup of tea, as if they had never ceased to be friends. I was impressed by the transforming power of getting hatred out of one's mind. That being said, I am not aware of any legal disposition based on this vision of original goodness.

Wolf: But this strategy of mediating reconciliation seems to be adopted by all cultures and not specifically related to contemplative practices.

Matthieu: Regarding the process of justice and punishment, the Dalai Lama has had interesting ideas about taking into consideration the extended suffering that a judgment might bring about. He once asked a panel of lawyers and judges in South America, "Two people have committed the same offense and, according to the law, they both deserve 15 years of jail. But one of them is a single father with five kids whose mother is no longer around, and the other one is a loner. If you punish the father by putting him in jail for 15 years, five innocent kids will suffer tremendously. So are you going to give the same sentence in both cases?" The lawyers said that it was indeed legally difficult to give a different sentence, but that judges would usually try to consider all the implications of their sentencing.

Wolf: Which implies that you do not solely judge and sentence the individual guilt but take the context into account with the goal of reducing suffering at a supraordinate level.

Matthieu: The purpose of justice is to diminish overall suffering, isn't it? Not just to take legal revenge. If justice is about preventing a criminal from doing harm again, with the idea of diminishing suffering for all, then you have to take into consideration the fact of creating possibly greater suffering by taking a person out of his or her life context. So ethics and justice should be less dogmatic and more embedded in human situations.

Wolf: Right. But this is difficult to realize in a canonical system based on the concept of free will, which attempts to assess the amount of subjective guilt and determines the severity of punishment accordingly. Your proposal requires a flexible legislation that allows the judge to consider not only the delinquent but also the consequences of the sentence for the society at large. This grants judges a lot of freedom and requires great confidence in their integrity and wisdom—and can of course easily become corrupted. Even if one left the evaluation of the wider context to a jury of 12 randomly selected people, the result could be highly biased by emotions.

Matthieu: It is true that although it would be desirable, this would be difficult to accomplish because it would be quite hard to formalize.

Wolf: Full agreement here. The law and procedures to protect norms should contribute to secure a more humane society. I like the idea of individually adapted, context-dependent judgments, but it comes with considerable costs—it requires consideration of many more variables and hence extended inquiry, and then it is exposed much more than our present practice to the personal opinion of the judges or the jury and the forces that might try to influence the judges' point of view.

CAN ONE PROVE FREE WILL?

Matthieu: Let's come back to free will with a crazy thought experiment. Imagine that you do something that does not make any sense, something that biological needs and normal, everyday brain computations have no reason to lead to. Let's imagine that I am sitting here feeling quite thirsty. I feel like taking a break and getting up to make a cup of tea and take a walk. When this wish comes in my mind, it has probably been brewing for a while in my brain. But now, instead of getting up to make a cup of tea, which would give me a kind of relief, I think, "Just to prove that free will exists, I shall sit here and stay put, even if I become terribly thirsty and hungry, pee in my pants, or faint." All the biological functions of my body are shouting to my brain, "Stop doing that! Get up, have a drink, go to the bathroom, and take a nap." Besides exercising free will, is there any kind of automatic computing in the brain that could lead me to do something that is so opposite to my natural needs? I could imagine all kinds of completely unlikely scenarios: rolling naked in the compost heap in the garden or buying an air ticket to Kazakhstan, Paraguay, or any place where I have no reason to go at the moment.

Wolf: If you really do such crazy things, your brain is more motivated to prove that it is free to decide than to seek comfort. If the decision is to stretch out on the garbage heap, your brain must be wired in a way that makes this demonstrative action more rewarding than all other alternatives.

Matthieu: Why would the brain be wired like that because this seems counterproductive for my survival?

Wolf: There must be some driving force. This can only be neuronal activity, and hence it must be generated by the brain.

Matthieu: Because feeling in control brings you some kind of a reward?

Wolf: Yes.

Matthieu: That sounds a bit extreme. Why would you want so strongly to prove that?

Wolf: A person who needs to do these aberrant things to prove to herself or others that she is autonomous, unbound, and free to decide is likely to have a problem with the experience of agency or authority.

Matthieu: Not necessarily. I would not act like that under any other circumstances. At this point, my only aim is to make the calm decision to convince someone that free will exists because I care enough about this philosophical question.

It reminds me of the story of the Nobel laureate Dr. Barry Marshal, who proved that the bacterium *Helicobacter pylori* was the cause of most peptic ulcers, not, as was widely believed until then, stress, spicy foods, and too much acid. Marshall was a bit discouraged that no one believed him and once said, "Everyone was against me, but I knew I was right." In the end, unbeknownst to his dear wife, Dr. Marshall drank a dense culture of the bacteria, expecting to develop, perhaps years later, an ulcer. He was astonished that only a few days later he became sick and developed acute gastritis. He said on BBC that he felt like he was drinking a disgusting cocktail as some sort of juvenile challenge. He was fully aware and dedicated and even had a sound sense of humor and did this crazy thing not because he was crazy but because he cared so much about the result for the welfare of humanity.

So, although contributing to resolving the question of free will may not save millions of lives, I might consider it worthwhile to behave in rather unconventional ways. I am sitting here in a state that is not completely deluded, hopefully, and I decide, if you say that this would be sound proof of free will, that it is worth sitting five hours in this chair.

Wolf: But you must have a pending problem, an internal urge that you want to resolve, if you do take on some crazy task like that, if you get up to roll around naked in the grass.

Matthieu: Maybe not if that would settle this philosophical issue that I feel is more important than looking like an idiot by rolling naked in the grass. I will not do so because I can't resist doing so out of an uncontrollable fit of madness but with a cool and clear mind for the sake of an argument that I think matters.

Wolf: Let's assume that you do this and you roll around in the grass. Will this be accepted as proof of free will?

Matthieu: That's what I am asking you.

Wolf: What was the precursor to this decision? What happened in your brain before the plan matured and the decision was reached? You would agree that planning and deciding must have happened in your brain.

In your example, it is the wish to prove to yourself or me that you have free will. So you have a concrete motive, and this motive has arisen from our conversation, from a conflict between my arguments and your feelings, a conflict that you want to resolve by demonstrating that you can decide on the fly to do something quite unexpected and prima facie useless. In this case, however, the antecedents of your seemingly free decision are clearly traceable. There were arguments questioning the existence of free will, these led to a conflict with your intuition, and then you thought about a way to resolve the conflict.

Matthieu: But here again, all you are saying is that nature abides with the laws of causality. The point here is about the factors involved in decision making. Is there room for a top-down causality originating from consciousness? We come back again and again to the fact that you cannot categorically discard the possibility for consciousness to be something other than a by-product of brain activity.

Wolf: We will discuss this more in depth in our next topic. But to return to our present focus, I think it is important to differentiate between creativity and free will. You showed creativity in imagining a possible scenario to solve the problem, but the decision to realize the action or to stay seated would again have depended on the actual state of your brain that in turn would have been set by a host of variables, some appearing in consciousness, some remaining in the unconscious. So what do you think would have happened in your brain if you had actually decided to do this crazy thing?

Matthieu: Crazy if I do it for no reason. But I have a sound reason. Suppose someone says, "Roll naked in the grass and we will spare the life of your kid." You would do it, and no one would think you are crazy. For me, clarifying the issue of free will—supposing that such an experiment would be logically compelling—makes it worth passing for a lunatic.

Wolf: So let's bring this down to the mechanisms. There must have been a pattern of neuronal activity that is the equivalent of this decision; otherwise nothing would have happened. So something must have brought about the pattern that is the substrate of this strange decision. What was in your mind? What would you think was the incentive, the *res movens*, the cause?

Matthieu: I intuitively feel that an element of consciousness is pushing it forward, and that in this case my respect for reason and wisdom makes it important for me to clarify the issue of free will.

Wolf: This intuition is at the origin of the fascinating question of mental causation, the question of whether mere thoughts or insights that appear in consciousness can influence future neuronal processes. This question is intimately related to theories on the nature of consciousness—a vast subject indeed, to which we should devote a separate discussion session, preferably after a good sleep and strong coffee, because this gets us to the limits of what we know and can imagine.

ARCHITECTS OF THE FUTURE

Matthieu:　On philosophical and logical levels, the question of free will is related to the larger one of determinism. Unless one goes down to the level of quantum physics, it seems obvious that events occur because of various preceding causes. At the gross level of our world, if things could happen without having any cause, there would be no laws of causality and anything could be born from anything: Flowers could pop up in the middle of the sky and darkness could be born from light. Without causal processes, our behaviors would be random and chaotic because no causal connection exists between our intentions and our actions. We certainly could not be held responsible for erratic or deviant behavior. There would also be no point in training the mind and trying to become a better human being because things would only happen randomly. This is indeed quite absurd.

On the other end of the spectrum, thinkers who support the idea of a strict determinism argue that if we were able to know perfectly the entire state of the universe at any particular moment in time, we could know exactly what has happened before and predict exactly what would happen next. Sentient beings would function like machines, and there would be no real choice in our actions. From birth to death, our lives and our selves would be entirely predetermined by the preceding states of the universe. If such an absolute determinism existed, any attempt at personal transformation would also be vain and illusory. We would be mere robots with the illusion of thinking and deciding things, when in fact we would never have any choice at all.

Pierre Laplace, for instance, was convinced that if an intelligence could know all the causes, conditions, and forces at work at a particular moment in the universe and had, in addition, the capacity to analyze all this information, "Nothing would be uncertain; the future and the past would be equally present to its eyes." But modern physicists have pointed out that this is impossible. As my friend and coauthor the cosmologist Trinh Xuan Thuan told me, "The uncertainty principle states that any measurement implies an exchange of energy. The shorter the time for the measurement, the more energy is needed. An instantaneous measurement would therefore require infinite energy, which is impossible. So the dream of knowing all the initial conditions with a perfect precision is mere delusion."[7]

Furthermore, strict determinism could only be possible if there were a finite number of factors involved in the state of the universe. However, if an unlimited number of factors is involved, including some that are probabilistic and including consciousness, and if all these mutually interact in an open system, such a system escapes absolute determinism.

Interdependence, a central Buddhist concept, refers to a coproduction in which impermanent phenomena condition one another mutually within an infinite network of dynamic causality, which can be innovative without being arbitrary and which transcends the two extremes of chance and determinism. It thus seems that free will can exist within such an unlimited network of causes and conditions that include consciousness. This brings to mind Karl Popper's logical argument according to which we can't predict our own actions because the prediction becomes one of the determinant causes that influence the action. If I predict that I'm going to collide with a tree in a given location in 10 minutes' time, this possibility makes me avoid the place in question, and so the prediction will not come true.

Wolf: I fully agree with the gist of what you say but would like to offer another reason for the unpredictability of the future that is at the same time compatible with determinism and with the idea that even if all initial conditions and all rules governing the system's dynamics were known it would still be impossible to predict its future trajectories. These seemingly contradictory ideas hold true for complex systems with nonlinear dynamics, and the brain is with all likelihood a highly nonlinear system. The reason is that such systems are highly susceptible to minimal perturbations. Thus, even if one had an exhaustive description of a brain's actual state and even if the processes leading from one state to the respective next followed the law of causality (i.e., were deterministic), which we believe is the case; even if we assume that there are no external intervening events, it would still be impossible to predict in which state the brain will be a couple of months later.

This is an established feature of complex systems with nonlinear dynamics, but it sounds counterintuitive to us because our cognitive systems generally assume linearity. Assuming linearity is a well-adapted heuristics because most of the dynamic processes that we have to cope with in daily life can be approximated with linear models from which we can derive useful predictions for appropriate reactions. Think of the kinetics of objects moving in the earth's gravitational field—a pendulum, for example. Once set in motion, its trajectory is nicely predictable. The same holds for a spear or a ball. However, if you take three pendulums and tie them together with rubber bands and set them in motion, their swings soon become completely unpredictable because of the complexity of the interactions among the three pendulums and the flexing rubber bands. Similar nonlinear dynamics are displayed by our financial, economic, and social systems because of the complex interactions between the agents participating in these densely interconnected networks. In these cases, our heuristics fail and give rise to sometimes severe misconceptions and false decisions. The problem here is that it is not only impossible to predict future trajectories (if it were otherwise, then everybody would win at the stock market), but it is also impossible in principle to control the future trajectory of the system by interfering with its variables. The recent crises of the financial market provide examples for this fact.

Matthieu: When it comes to inner phenomena or mind events, the impossibility of knowing all initial conditions to make a prediction about future mental states becomes even clearer. Take, for instance, the knowledge of the "present moment": The moment you know the present moment, it is no longer the present.

According to the Buddhist view, our thoughts and actions are conditioned by our present state of ignorance and the habitual tendencies that we have accumulated in the past. But wisdom and knowledge can put an end to ignorance, and training can erode past tendencies. So, ultimately, only someone who has achieved perfect inner freedom and full enlightenment can truly have free will. It is said that the Buddha is free from past karmic influences. He has exhausted the maturation of past negative action, so what he does is purely an expression of his inner wisdom and compassion. A Buddha is totally alert to the minutest aspects of cause and effect. Consider Guru Padmasambhava's famous saying: "The view can be high as the sky, while our careful understanding of the law of cause and effect should be as fine as flour powder." The more you understand emptiness, the more you understand transparently the laws of cause and effect.

Freedom from conditioning could be the essence of free will. An enlightened being acts appropriately according to the cause and needs of everyone and is not influenced by past tendencies. It also seems that even before achieving the ultimate goal of enlightenment, when someone is able to remain for a few moments in the limpid freshness of the present moment, a state of pure awareness on which rumination on the past and anticipation of the future have no bearing, this should be a state conducive to the expression of free will.

Beings who are still under the yoke of delusion act according to the force of past habitual tendencies. According to the Buddhist view, people who have some incentive to kill or hate others have built up that propensity not only earlier in this life but also in past lives. Yet even if we are the product of the past, we are still the architects of our future.

Wolf: At this point, we clearly enter the realm of belief and metaphysics; because these dimensions are orthogonal to science, neurobiology can neither support nor falsify any of these claims.

THE NATURE OF CONSCIOUSNESS

SOMETHING RATHER THAN NOTHING

Consciousness is fundamentally a fact of experience. But where does it come from? Can consciousness be reduced to brain activity or should it be seen as a "primary fact" that can only be understood by direct experience, preceding all other experience and knowledge? How is the "first-person" approach of people who meditate and phenomenologists different from the vision according to which consciousness is a phenomenon resulting from neuronal interaction? Can we envisage an intermediary that takes into account the brain's place within the body, itself placed within a society and a culture? What to think of parapsychological phenomena?

Matthieu: In Buddhism we consider consciousness to be a primary fact. Let me explain what we mean by that. When you investigate the material world, as you refine your analysis, you eventually come down to atoms, then elementary particles, quarks, or superstrings, and finally to the quantum vacuum—in short, to the most fundamental aspect of matter. Once you are there, Leibniz's question was, "Why is there something, rather than nothing?" Unless you bring into the picture a completely different kind of explanation, such as a creator, you can't really answer that question. You simply have to acknowledge the presence of phenomena. We then have to consider that matter or the inanimate phenomenal world is a primary phenomenon. From there, we can describe these phenomena, try to understand the nature of their most basic constituents through classical and quantum physics, and study how they build up the visible world, but we can't answer the question of why there is something rather than nothing. It just happens to be there rather than not there. We can, however, investigate further the nature of this phenomenal world. Is it endowed with intrinsic existence, as realists would contend, or is it simply manifesting while being void of intrinsic, solid existence, as Buddhist philosophy and quantum physics would say?

We can make a similar investigation into the nature of consciousness. But we need to do so in a coherent way and keep going until we reach its most fundamental aspect. What do I mean by "coherent way"? There are two main methods of approaching consciousness: studying it from the outside (the third-person perspective) or studying it from the inside (the first-person perspective). By "outside" and "third-person" perspectives, I mean the study of the correlate of conscious phenomena in the brain, the nervous system, and our behavior as it can be observed by a third person, who does not experience what I experience. By "inside," I mean the actual experience.

We can of course provide descriptions and explanations about the way consciousness arises from the increased complexity of life and the building of a nervous system and a highly developed brain. We can correlate thoughts, emotions, and other mental events with various kinds of brain activity, and we know in an increasingly refined way how various cognitive and emotional processes are related to specific areas of the brain. This is the third-person analysis. We can also question someone in great detail on how he can describe what he feels. This is also called the "second-person perspective" because it is achieved through an interaction with someone who is helping you to describe your experience in depth.

But truly, without subjective experience that we can apprehend introspectively, we could not even talk about consciousness. We would not be able to talk about anything in fact. This experience can never be truly and fully described from a third-person perspective. What is it to feel love? To experience seeing the color red? You might write a thousand-page description of what is going on in someone's brain, in his physiology, and his outer behavior; although you might learn a lot in this way, you will never know exactly what love or seeing red is to him unless you have made a similar experience yourself. If you haven't tasted wild honey, then no amount of description will allow you to experience its sweetness for yourself. Buddhist scriptures tell the story of two blind men who wanted to understand what colors were. One of them was told that white was the color of snow. He took a handful of snow and concluded that white was cold. The other blind man was told that white was the color of swans. He heard a swan flying overhead and concluded that white went *swish swish*.

So, once again, "consciousness" would not really mean anything without subjective experience. Therefore, to be coherent, we must fully pursue its investigation from that perspective, without constantly jumping from outside in and from inside out as we please. We must follow a consistent line of investigation until its ultimate point.

What happens if you go deeper and deeper into experience? Where do you end up? From that first-person perspective, you will never come to neurons. As you know, I have participated in neuroscience studies on the effect of meditation on brain functions, and I have seen on the screen how meditating on compassion will activate the anterior insula. But, subjectively, forget about brain area localizations—unless I have a strong headache, I don't even feel or experience that I have a brain.

Wolf: This is true. We have no recollection of the neuronal processes in our brain. They are transparent. In the case of a headache, the signals causing pain come from the meninges. The brain is pain-insensitive.

Matthieu: I could also say that when we are engrossed in thoughts and perceptions, the naked experience of awareness devoid of mental constructs is transparent to us. Yet I end up here when I refine and pursue further and further my investigation of subjective experience: pure awareness, basic consciousness, the most fundamental aspect of cognition. This basic awareness does not necessarily need to have particular content, in terms of mental constructs, discursive thoughts, or emotions. It is just pure awareness. As I described earlier, this is sometimes called the "luminous" aspect of mind because it allows me to be aware of both the external world and my internal world. It allows me to recall past events, envision the future, and be aware of the present moment.

But when you come to the most refined state of awareness, which is devoid of any particular content except for its own lucidity and clarity, if you ask again, "Why is it there rather than not there?", here too you have to say, "It is simply there, and I can only acknowledge it." We may thus consider that, here too, we are dealing with a primary phenomenon. From the experiential, phenomenological point of view, pure awareness precedes everything, whether being aware that I am alive or postulating a theory about consciousness.

The main point I want to emphasize here is that we need to diligently follow one line of investigation until its end. In physics, when you investigate phenomena from the perspective of quantum mechanics, the notion of solid reality as being made of independent particles soon loses its meaning. Yet you need to pursue this analysis to its end to arrive at this conclusion. You cannot continually jump back and forth from quantum mechanics to Newtonian mechanics at your convenience simply because you want to hold onto a realistic explanation of the phenomenal world.

Just a remark also about treating consciousness as a mere phenomenon: From the subjective point of view, phenomenologists would say that consciousness is nothing more than the appearance of phenomena—the global acknowledgment that there is a world out there and the specific awareness of there being a multitude of varied phenomena. For them, consciousness is not just a phenomenon among others that can be studied like an object because whatever you do or whatever question you study, you cannot step out of consciousness.

Wolf: This is truly a major epistemic problem. Apparently we have a phenomenon that does not allow us a coherent, unified interpretation. Because, as you said, if we approach the issue from the third-person perspective and analyze matter, we will never encounter consciousness. We will also not encounter consciousness when we analyze the brain. We see spatiotemporal patterns of neuronal activity, we see particular states, if we refine the analysis we see electrochemical processes, and all the way down until we finally see molecular processes, but we will never encounter anything that resembles our experience of being conscious. The same of course is true for all other behavioral manifestations. If we follow the flow of activity through the brain, starting at the sensory surface of the body and proceeding along the sensory pathways until we finally leave the brain on the motor side, we will never encounter the experience that emerged from the observed activity. We have seen activation patterns, we have seen them change in correlation with sensory stimuli and motor responses, but we have not encountered what the brain experiences from its first-person perspective.

Matthieu: Well, you can't really say that. All you can say is that we, as first-person subjects, experience all kinds of things, including the finding that there is a brain when we open the skull. You cannot lend to the brain considered as an object of knowledge the faculty of knowing as a subject does, without mixing up the third- and first-person perspectives, which are mutually exclusive. All you can say is that you experience things. You have no idea about what I experience, and you can't know for sure either, by examining it, whether an object like the brain could ever experience something.

Wolf: Still, we are able to establish links between phenomena experienced from the first-person perspective and processes observed from the third-person perspective. Take pain as an example. The subject experiencing pain and a third person observing the subject in pain can reach a consensus on what it is to experience pain. Thus, first- and third-person perspectives can be correlated. In addition, it is possible to operationalize the feeling of pain through subjective ratings of intensity and quality on standardized scales. These parametrized data can in turn be correlated directly with activity in specific regions of the brain. Conversely, lesions in pain-mediating pathways or pharmacological disruption of signal transmission can abolish the sensation of pain. Given these close correlations between subjective feelings and the underlying brain processes, I wonder whether the epistemic distinction between first- and third-person perspectives is actually as much of an insurmountable problem as we thought.

The fact that every individual attaches her own unique connotations to the contents of awareness, be it feelings, intentions, or beliefs, is close to trivial, given the relative uniqueness of the contexts in which these experiences are embedded and conceptualized. Within the range of genetic variability, we all have more or less similar receptors mediating the sensation of cold. However, each of us learned to attach the word *cold* to a particular sensation in a different context. Hence, the field of associations in which the word *cold* is embedded differs from individual to individual, as does the feeling described with the word *cold*. The same applies for all connotations of conscious experience, and even more so for more abstract concepts such as intentionality and responsibility, because these depend much more on cultural conventions than concepts such as cold. Still, all experiences, whether evoked by a physical stimulus or social interaction, are mediated by brain processes. If people conclude that a given condition evokes consistent experiences, they usually coin a term for the respective experience. Henceforth, this experience assumes the status of a social reality, of an immaterial object, of a concept on which different subjects can focus their shared attention.

As you say, if we take the first-person perspective as a source of insight into brain processes, we are aware of perceptions, decisions, thoughts, plans, intentions, and acts; we experience these as ours; and we can even be aware of being aware and communicate this fact. You also mentioned that experienced meditators can apparently cultivate this meta-awareness to the point where they are aware of being aware without requiring any concrete content of that awareness. We should, at a later stage, try to explore more closely the neuronal processes that might underlie this special state. For the moment, I would just like to point out that none of these first-person experiences tells us anything about the neuronal processes from which they emerge. We are not aware of neurons, electrical discharges, or released chemical transmitters. Accordingly, the seat of the conscious mind was thought to be in many places all over the body before scientific investigations made it clear that it should be sought in the brain.

Matthieu: Some people think that the mind is in the heart.

Wolf: Yes, in situations where strong emotions make our heart beat faster or make us feel a weight in our chest, we feel the somatic manifestations of a mental conflict, but both the conflict and the perception of its bodily manifestations are based on brain processes.

Matthieu: You agree that by keeping consistently with the third-person approach, you will never encounter consciousness. To do so, you have to leave the third-person perspective and adopt the first-person perspective. Therefore, how can you hope to prove that consciousness is reducible to the brain when you accept the idea that you can't encounter it by analyzing matter, and you need to use your own subjective experience to articulate your theory? You conclude that consciousness is a byproduct of the brain, but your reasoning process relies on your first-person experience, and by doing so you are tacitly treating your own experience of consciousness as a primary fact.

Wolf: As I tried to argue earlier, neuroscience can explore the mechanisms underlying the cognitive functions required for the emergence of consciousness (i.e., the mechanisms developed during biological evolution and refined by epigenetic shaping). Because animals are conscious and have experiences and feelings that can be operationalized, much of the relevant research can be performed on animals. Complications arise when we try to account for the unique connotations that humans have attributed to consciousness and experiences. These attributions are the products of cultural evolution and are conceptualized in our symbolic language systems. Humans are embedded in a dimension of social realities that they have created by interacting with one another, observing one another, and sharing their observations and subjective experiences.

Through these communicative processes, humans exchange descriptions of their first-person experiences, establish consensus about the congruency of these experiences by attaching labels (names or symbols) to these experiences, and assure each other that these experiences are common to all human beings. In this way, these immaterial phenomena, accessible only from the first-person perspective, gradually acquire the status of realities that one can talk about, attribute to others, and integrate into one's own self-model. Hence, many of the qualities that we associate with first-person phenomena are actually self-attributions drawn from collective experience and manifested in concepts for which we have coined linguistic terms. These attributions emerged from cultural interactions and therefore defy neurobiological explanations that are confined to the analysis of individual brains. However, an epistemic bridge between these first-person phenomena and neuroscientific approaches still exists. The first-person phenomena and their descriptions owe their status to social interactions among agents who are endowed with the specific cognitive functions of human beings, which are amenable to neuroscientific analysis.

The emergence of new qualities and the need to build bridges between different levels of description is actually common practice in scientific disciplines that deal with complex systems. Here is an example from neuroscience: Behavior emerges from the complex interactions among sensors, neuronal networks, and effector organs. To describe and study behavior, one has to apply the tools and description systems of the behavioral sciences and psychology, whereas completely different descriptions and methods of analysis are required for the study of the neuronal underpinnings. Still, correlations and, in lucky cases, even causal relations can be established between the explananda defined in the different description systems. If the cultural realities that define the mental or spiritual dimension of our existence have emerged from the complex social interactions of human agents, and I am convinced that they have, then similar bridges should exist between description systems that deal with mental phenomena and those that refer to sociocultural processes. Finally, the latter should in turn be relatable to processes taking place in human networks. In a highly simplified way, one could say that interactions between neurons lead to behavior and cognitive functions, whereas interactions between cognitive agents—in this case, human subjects—lead to social realities.

Let us briefly return to the curious fact that we are unable to experience the brain processes that bring forth our cognitive abilities. We have no feeling for the neuronal machinery that prepares our experiences, interprets sensory signals, and presents the respective reconstructions to our consciousness. The question is, Who are "we"? Who is the observer in this game? Again, introspection and scientific evidence suggest radically different answers probably because we are only aware of the results of our brain processes and not of the processes themselves.

Matthieu: One could reply to this argument that the neuroscientist is not aware that his perception of and experimentation on the brain, as well as his interpretation of his observations, presuppose consciousness.

Wolf: We encountered this problem before when we talked about the self and were confronted with the discrepancy between the ordinary intuition that it should reside in a privileged single site in the brain and the neuroscientific evidence that there is no such site. Apparently, the two sources of knowledge, introspection and science, provide different answers. For a long time, these two systems seemed so far apart that bridging their theories appeared out of reach.

Matthieu: There still might be a bridge between the two perspectives if we follow the position proposed by Francisco Valera, who often said that even the knowledge derived from the third person is in fact a byproduct of working on a host of first-person experiences. For instance, you can extract from the first-person experiences that are shared by a group of subjects some invariants and some structures, such as mathematics and the laws of physics, which govern phenomena.

Wolf: Probably the best that neurobiology can do at the moment is to list the neuronal processes that need to be functional to support what we subjectively experience as consciousness, whereby the definition of consciousness is for now an operational one. We contrast consciousness with not being conscious, with being comatose or in deep sleep. We can list a whole number of mechanisms that need to be intact to support a brain state that is capable of manifesting consciousness. We know that if we inject anesthetics, we can change these brain states, and that certain brain states lead to a loss of consciousness. Also, consciousness is not a static phenomenon. One can be brightly awake and highly alert, but one can also be dozing off, distracted, or less bound in the present. So consciousness may be a graded phenomenon.

Matthieu: Buddhism speaks of six, seven, or eight aspects of consciousness. It speaks first of the ground or basic consciousness, which has a global, general knowledge that the world is there and that I exist. Then there are five aspects related to the five sensory experiences: seeing, hearing, smelling, tasting, and touching. The seventh aspect is mental consciousness, which associates abstract concepts to the first six aspects. Sometimes there is considered to be an eighth aspect of consciousness that is related to afflictive mental states that distort reality (hatred, craving, etc.). But even more fundamental than all these states and aspects is primary consciousness, what is called the continuum of the luminous fundamental consciousness.

In Buddhism, the matter/consciousness duality, the so-called mind-body problem, is a false problem given that neither of them has an intrinsic, independent existence. According to some Buddhist teachings that analyze phenomena at a more contemplative level, the primordial nature of phenomena transcends notions of subject and object or time and space. But when the world of phenomena emerges from primordial nature, we lose sight of this unity and make a false distinction between consciousness and the world. This separation between the self and the non-self then becomes fixed, and the world of ignorance, samsara, is born. The birth of samsara did not happen at a particular moment in time. It simply reflects at each instant, and for each of our thoughts, how ignorance reifies the world.

Buddhism's conception is thus radically different from Cartesian dualism, which postulates on one side a truly existing solid material reality and, on the other side, a completely immaterial consciousness, which cannot have any real connection with matter. The Buddhist analysis of phenomena recognizes the lack of intrinsic reality of all phenomena. Whether animate or inanimate, they are equally devoid of autonomous, ultimate existence. Thus, a merely conventional difference exists between matter and consciousness.

Because Buddhism refutes the ultimate reality of phenomena, it also refutes the idea that consciousness is independent and exists inherently, just as much as it refutes that matter is independent and exists inherently. This fundamental level of consciousness and the world of apparent phenomena are linked by interdependence, and together they form our perceived world, the one we experience in our lives. Dualism lacks the concept of interdependence and postulates a strict separation between mind and matter; Buddhism states that emptiness is form and form is emptiness. Accordingly, the dichotomy of "material" and "immaterial" makes no sense.

In other words, Buddhism says that the distinction between the interior world of thought and the exterior physical reality is a mere illusion. There's only one reality or, rather, only one *lack* of intrinsic reality! Buddhism does not adopt a purely idealist point of view or argue that the outer world is a fabrication of consciousness. It just points to the fact that without consciousness, one cannot claim that the world exists because that statement already implies the presence of a consciousness.

This might sound puzzling, but it resembles the answer given by some cosmologists when asked what was there *before* the Big Bang. They say that this question does not make sense because time and space began *with* the Big Bang. Likewise, anything we can ever say about the world, the brain, and even consciousness begins with consciousness. Even the question, "But couldn't a world totally deprived of life and sentience exist on its own?" as well as any answer that you might like to give to this question—all of this presupposes consciousness. Of course, it would be foolish to deny the existence of lifeless worlds because most planets are indeed lifeless, but without consciousness, in a way, there is no question, no answer, no concepts, no "world" as an object of experience.

It seems that we never place ourselves "outside" consciousness, even when we try to determine its nature and origin. This argument resembles Gödel's second incompleteness theorem, which says that mathematical theories cannot demonstrate their own consistency, which can also be understood more generally as saying that we are always limited in our knowledge of any system when we are part of that system.

Wolf: Let me comment on this epistemic circle with a thought experiment. Imagine *homo sapiens* did not evolve—a scenario that is not unlikely given what we know about the unpredictable course of evolution. There would be no culture, no language, and no conceptual framing of observable phenomena. However, unless you take the radical position that there would be no universe, no planet earth, and no organisms without human beings observing and describing all these phenomena, there would still be a host of living organisms, most likely including nonhuman primates. They would have feelings and experiences that they could remember, and they would be conscious during two-thirds of each day. The only radical difference would be that none of the organisms would be able to become aware of participating in an immaterial dimension of the world characterized by mental constructs and attributions because for them this dimension would not exist—indeed, the concept of such a dimension would not exist. Even in our cultural world created by humans, animals only marginally participate in this world of mental constructs because they lack the cognitive abilities to experience them. I say "marginally" because domesticated animals such as dogs are to some extent able to participate in some aspects of the social realities constructed by humans. For example, they are capable of understanding gestures that indicate where they should direct their attention, a function addressed as shared attention.

Let us then return to the phenomenon that we address as consciousness, the ability to not only have an experience or a feeling but to be aware of it. We probably need to distinguish between consciousness as such and the state that allows one to be conscious of something. The latter can vary substantially because brain states are graded. Also, many brain processes bypass consciousness and lead to action without us being conscious of the causes that triggered the action. As we discussed already when addressing the differences between conscious and subconscious processing, many subconscious or unconscious processes go on while a person is conscious. It is commonly held that one can only be conscious or aware of something if it is attended to, if the focus of attention is on this particular content. These contents can be sensory signals from the outer world or from the body, or processes generated in the brain, such as emotions, inner states, and feelings. The focus of attention can be shifted either intentionally, in a top-down fashion, or in a bottom-up way by salient external stimuli—the sudden appearance of an object or a sudden change in the environment automatically attracts one's attention. Thus, attention is one of the mechanisms required to bring a content into consciousness, suggesting that there is a threshold for contents to reach consciousness. Furthermore, as discussed previously, the capacity of the workspace of consciousness is limited. Finally, we equate conscious processing with reportability. If subjects are unable to explicitly recall or report on an event, then we assume that the respective information has either not been processed at all or processed only in the subconscious. To what extent, however, do these insights help us define the essence of consciousness?

Matthieu: Is it not the most basic faculty of being aware? The content changes all the time. Any number of studies can be focused on the content of experience, on how much information we can hold in the mind, on the mechanisms of sensory perception, and on how memory can influence what we see and what we hear. Ultimately, the most fascinating question is, "What is the nature of this most basic faculty of knowing?"

Wolf: What could it be that it makes us aware of ourselves? Let us talk about the different levels of awareness. First, I think one of the most basic levels is phenomenal awareness, the ability to be simply aware of something. Then comes the ability to be aware that you are aware of something. Finally, there are the more self-related aspects of consciousness: One is aware of being an individual who is autonomous, capable of intentional acts, and separate from other individuals. One is also aware of one's conscious self, which is probably the highest level of metacognition. To me it seems that these higher forms of consciousness are absent in the animal kingdom and result from experiences and self-attributions of phenomena that owe their existence to cultural evolution and whose dimension can only be experienced by cognitive systems as differentiated as the human brain.

Matthieu: I would call it *self-illuminating awareness* instead. The expression *conscious self* could easily be misunderstood as assuming the existence of an autonomous self at the core of ourselves. We discussed that earlier. As for animals, a significant number of species—great apes, elephants, dolphins, and magpies, for instance—has passed successfully the "mirror self-recognition test," which shows that they do have a sense of self and recognized themselves in a mirror. Human children pass this test when they become 18 to 24 months old.

Wolf: The point I want to come to is that consciousness—or the aspect of consciousness that creates such big problems in epistemology—is a construction that results from an interpersonal discourse. It is a social reality, much like the construct of free will, and therefore has a special ontological status. We will probably see this more clearly if we consider consciousness in animals in contrast to consciousness in human beings. My hypothesis is that the epistemic problem arises from the fact that we do not consider sufficiently that brains, our cognitive organs, are embedded in bodies, and that this ensemble, the person or individuum, constitutes one node in a complex network of interacting persons—and that the self-model of these persons is only what it is because of their embedding in a society of similar agents who interact with and mirror each other. Through this interchange, new phenomena are brought into the world that would not exist had there been only one person, one brain.

Matthieu: Is this similar to what Francisco Varela called the *embodied mind* or *enaction*, the fact that consciousness from his point of view is the brain in a body in an environment, and you cannot dissociate those three?

Wolf: These social realities are abstract concepts created because human beings entered into dialogues, telling each other that they are able to imagine what it is to be the respective other, to imagine that this respective other has certain feelings and aspirations, that they share certain forms of reasoning and can also share their attention by looking at the same object. Through this reciprocal discourse, attributes such as consciousness and free will have been conceptualized, attributes that the respective other would not be aware of had he or she grown up alone. My hypothesis is that through this social interaction, which is at the origin of cultural evolution, realities have been created that can readily be experienced as such but that transcend the reality that existed before cultural evolution began. These realities exist in the "in between." The objects in this dimension of reality are relational constructs; they are immaterial, not tangible, not visible, and not directly accessible to our senses.

Examples are values and beliefs, confidence and justice, intentionality and responsibility, and the various attributes of consciousness. All these social realities have a special ontological status that differs from both the material world and that of the precultural biosphere.

The phenomenon that we address with the term *consciousness* would not exist without this dialogue among human minds, without education, without the embedding in a rich sociocultural environment, and without the mutual attribution of mental constructs. These constructs are internalized and become implicit properties of our selves. We experience them as part of our reality and invent terms to name and describe them. They are similar to values. These, too, are social constructs and are not found in the brain; all we can do is identify systems that assign value to certain brain states and couple them with emotions. The same holds for all the characteristics that one associates with consciousness. We do not find consciousness in brains, but we can try to identify structures that are necessary for consciousness to manifest itself.

Matthieu: All of what you describe—and you spoke about various levels of consciousness—would correspond quite well to what Buddhism calls the "coarse" aspect of consciousness, which is the aspect of consciousness involved with the complex world of information, perceptions and their interpretation, relating things with each other, and feeling emotions in reaction to outer events or inner recollections. None of these would ever arise without our constant interaction with the environment and other sentient beings. The body is embedded in the universe and has evolved to process this embodiment in the world in the best possible way to efficiently interpret all stimuli and relate in a coherent way to the world and others. The process of evolution has allowed the appearance of this incredibly efficient method of integration in the world.

But we still have to understand the most fundamental aspect of "pure awareness," what Buddhism refers to as the "subtle" aspect of consciousness. Recall the example I gave earlier about a beam of light—it reveals what is there without being modified by it. Likewise, according to Buddhist contemplatives, pure awareness is neither obscured nor modified by the content of thoughts; it is unqualified and unaltered.

CULTIVATING STATES OF SUBTLE CONSCIOUSNESS OR PURE AWARENESS

Wolf: I think what you call *pure awareness* might correspond to a state of "solution"—as I proposed earlier—a state where the brain is without conflict and not searching for the answer to a question or attempting to solve a problem. Once the brain is in such a "eureka" state, special subsystems become active and accomplish three tasks: mediate the feeling of satisfaction associated with this state of mind temporarily devoid of inner conflicts. enable learning to take place (a solution state is a good condition in which to learn because ambiguity is low), and eventually terminate this " pleasant" state to prepare the brain for the processing of new information and the search for the next solution.

Here is my question. Could it be that you create a state of high arousal, an attentive, highly awake state, and then use all your attentional resources to maintain this state without selecting particular contents for processing? The difficult part seems to be to prepare a workspace for conscious processing without filling it with concrete contents. Normally, and this feature is a characteristic of the workspace of consciousness, different contents can be represented simultaneously, and they are semantically bound together. This is why we talk about the unity of consciousness. Interestingly, these binding functions actually seem to involve high-frequency oscillations and synchronization of oscillatory activity across large distances in the brain. However, to prepare a clear state that is sustained but not filled with content, don't you have to achieve a double task? You have to prepare the workspace, which requires investing attentional resources, and at the same time you have to use your attentional resources to protect the workspace from intrusion by wandering thoughts rather than select contents to be represented.

Matthieu: Once you have become familiar with this process, it can become effortless and uncontrived. You are not trying to prevent anything from arising, but when something presents itself, you just let it come and let it go so that it doesn't make waves.

Wolf: So you don't pay particular attention to the coming and going of contents, you just let them happen—

Matthieu: You don't try to stop it, but you don't encourage it either.

Wolf: But how can you then prevent this internal chattering from taking place if you don't repress it actively?

Matthieu: Because internal chattering is usually due to the proliferation of simple thoughts. Without repressing them, you can simply let them vanish as they arise. There is no point trying to stop perceiving the outer world, hearing the birds that are singing right now outside our window. You simply let thoughts and perceptions arise and undo themselves. In Buddhist teachings, one gives the example of drawing with a finger on the surface of a lake. If you draw the letter A, it vanishes as you draw it. It's not like engraving it in stone. Another example is a bird passing through the sky without leaving any trace. There is no point trying to prevent thoughts that are already there. But you can certainly prevent them from invading your mind.

Wolf: Could one say that you use your attentional resources by directing them to the preparation of this internal workspace while taking care that external or internal intrusions—sensory signals or thoughts and emotions—can pass or enter that space but do not stick there?

Matthieu: That's right. If you don't stick to them, they don't take hold in the system.

Wolf: They don't take hold in or dominate the system. You keep the system as clean as possible by not paying much attention to these intrusions and letting them come and go. You free the workspace from randomly intruding contents, and then you fill it with selected contents that you call on through intention—for example, empathy or compassion—granting them a privileged space. Is this what you do when you meditate?

Matthieu: That's the essence of meditation. It means, as we mentioned earlier, becoming familiar with something and cultivating a skill in a methodic, nonchaotic way. It is not a semipassive way of learning but a fully engaged one, conducted in a coherent way.

Wolf: In other words, you prepare a workspace at a metalevel of awareness that allows for the generation of unified states. Then through deliberate selection of contents, you generate a fairly homogenous, uncontaminated internal state that is devoid of competitive interactions. Subsequently, through repetition of this process, the representation of the selected state is consolidated. This reminds me of the stages of a typical learning process: Be alert, be prepared to process information, prepare your processing stages properly by being quiet and not distracted, concentrate on the particular content that you want to learn, and rehearse it. Once the content is safely stored and retrievable, the trainee becomes an expert.

Matthieu: The essence of cultivating the mind is indeed familiarization and sustained training. When you are not overly influenced by past memories or invaded by thoughts on future developments, you can remain in clear awareness, in the freshness of the present moment. This will allow you, when you want to do so, to cultivate fundamental human qualities such as altruistic love and compassion, but also to think lucidly about the past and future when needed. It does not mean, of course, that you will be "stuck" in the present moment and become dysfunctional, as some might fear.

Wolf: Recent experiments indicate that the conscious state is characterized by large-scale coordination of neuronal activity across many different cortical areas. With all likelihood, this coordination is realized by entraining large ensembles of neurons into coherent activity. One effect of such coherent states is that signals can be exchanged particularly effectively and rapidly between the distributed processing modules. This would be the basis for binding together distributed results into unified concepts. The fact that attention is so important in this process is compatible with the notion that only contents that are the focus of attention enter consciousness, and that attention creates states of increased coherence. If one selects visual signals for further processing, the visual centers get entrained into coherent oscillatory activity. This renders visual signals more synchronous and, as a consequence, more salient, which in turn increases the likelihood of their entering into consciousness. Thus, a conscious state can be considered a dynamic state where large networks of processing areas engage in coherent activity.

This coherence could in turn provide the temporal frame for the unifying representation of distributed computational results. Preparing that workspace of awareness could be equivalent to engaging large neuronal networks in coherent activity. At the beginning of a contemplative exercise, you prepare the workspace but don't fill it intentionally yet, and contents will enter and leave, loosely associated like in a dream, without being allowed to stabilize. Apparently, it is then possible to intentionally select particular contents, bring them into this empty workspace where they can evolve without much interference, and sustain them. If this were a state of highly coherent activity, it would be ideally suited to stabilize itself through learning. Studies on neuronal plasticity show clearly that coherent or synchronized activity, if maintained over a sufficiently long period of time, will lead to changes of synaptic connections in a way that stabilizes the coherent state and later on facilitates reproduction of this state.

Matthieu: We use a similar image, saying, "Let compassion fill the entire space of your mental landscape."

Wolf: Exactly! A single content takes the whole coding space and is sustained dynamically until it is eventually engraved by learning and becomes a behavioral disposition like an expert skill, automatized....

Matthieu: We would prefer to call it natural or uncontrived because familiarization with pure awareness brings about freedom from automatic thinking and habitual tendencies.

This process is how compassion, for instance, can become "second nature." You embody compassion. This notion refers to a genuine long-term transformation, not just a flash of experience, a flash of compassion that will not last. That's the heart of the spiritual path.

Wolf: This example illustrates one of the important functions of the interplay between attentional mechanisms and consciousness. Being conscious allows one to select which of the many possible contents should be bound together into coherent wholes, into the unified constructs that characterize conscious experience. As you suggest, one can subject to this intentional consolidation process not only experiences mediated by sensory systems but also self-generated, internal states such as emotional dispositions, which after significant practice would become natural behavioral or cognitive dispositions.

Matthieu: I would prefer to say that such a skill can become consummate and effortless. When you know how to ski well, you can glide downhill without always being tense and worried about falling down.

Wolf: The point is that it is, as you say, effortless. This is also what qualifies automatic processes. They do not require investment of attention or conscious recall of instructions or strategies.

Matthieu: We should be a bit careful with "automatic" mental process, which often perpetuate deluded habitual patterns and perceptions. Let's say that they don't require forceful attention. So you are not effortfully attentive, but you are not distracted at the same time.

ON VARIOUS LEVELS OF CONSCIOUSNESS

Wolf: I would like to briefly return to the question of different levels of consciousness, in particular to the question of content-free or empty states of awareness.

Matthieu: Empty in the sense of being free from content, but not empty in the sense of perfect clarity. It is a state of extreme awareness of its own clarity. Light can shine on a dark sky, and yet it doesn't light up anything in particular.

Wolf: You can be aware of the fact that you have a platform that allows you to be aware of content without any content being there.

Matthieu: That we call *nondual consciousness* because there is no separation between subject and object.

Wolf: Right. I think I would address this as *meta-awareness*, the awareness of being aware or conscious. If a content then appears, one becomes aware of that, but from the perspective of being an observer of one's own consciousness.

Matthieu: There is pure awareness. But there is no split between a subject that knows something and an object that is known.

Wolf: For this to happen, the "observer" would again have to fuse with the level of consciousness at which the content is represented. Perhaps this is a second step.

Matthieu: That is what we call *dwelling in nondual awareness*, and that is the most fundamental form of experience. So what is the nature of that most primal pure awareness? That's really the question.

Wolf: Indeed! What could have been the evolutionary processes, the selection pressure, that brought forth brains that are capable of generating a space for the unified, conscious representation of contents, and then of being able to be aware of having this representational space? Maybe it is helpful in this context to recapitulate what we know about the evolution of brains. In brains of lower vertebrates, a fairly short path exists from cortical areas that process the signals of sensory organs to the executive areas that program the reaction. These relatively short sensorimotor loops are of course much more elaborate than simple reflex loops because signals are highly processed and transmission is conditional on past experience and input from other systems.

What distinguishes the more evolved brains is essentially the addition of new cortical areas, whose intrinsic organization is, however, strikingly similar to that of cortical areas in less evolved brains. The main difference is the way in which these new areas are embedded in the already existing networks. The areas added at late stages of evolution do not communicate with the periphery, do not get direct input from the sensory organs, and do not have direct connections to the effector organs or muscles. Instead, they are mainly connected with the cortical areas that evolved earlier.

This principle is maintained throughout the whole evolutionary process. More and more cortical areas are added that talk to each other. We believe that the different cortical areas perform similar computations because their internal circuitry is so similar. In the brain, function is determined by the architecture of connectivity, and hence similar architectures are likely to support similar functions. These purely anatomical considerations suggest that the evolutionarily recent areas process the results of the older areas in a similar way as the older areas process signals from the outer world. This iteration of cognitive processes across several hierarchical levels could thus generate representations of representations (i.e., metarepresentations). Information that has already been processed becomes the object of yet another cortical computation (i.e., of a secondary cognitive operation). These iterative operations can even be circular because most of these areas are interconnected. In principle, this permits the generation of metarepresentations of increasingly higher order. In other words, highly evolved brains can apply their cognitive functions to the outer world and processes that occur within the brain. Brain processes become the object of the brain's own cognitive operations. This could be the basis of phenomenal awareness, the awareness to perceive something, to run a protocol of what one perceives, and, in the case of human beings, to talk about it. Animals probably show some of these abilities because their brains are organized in a similar way.

Curiously, however, this awareness of internal processes provides no clues as to the computational operations underlying these cognitive functions. We have no insight into the neuronal processes that bring about cognition. We are only aware of the results—just as we are aware of an action without being able to tell which neuronal processes in the motor centers of our brains caused this action.

Another intriguing question in consciousness research is whether having evolved a platform for conscious processing has any survival value. What would happen if the human brain functioned without being conscious of its functions? Would that make a difference? The philosopher David Chalmers says it wouldn't. He argues that consciousness is just an epiphenomenon, and we would do just as well without it because the underlying brain processes would be the same and get us through life without us having to be aware of them. I doubt that this is the case. I believe that being aware of one's cognition and being able to communicate with a symbolic language system contributes to the understanding of others, the generation of social systems, and eventually the development of differentiated cultures. These abilities increase fitness because they allow individuals to cooperate, refine models of the world by sharing and comparing experiences, and develop diversified coping strategies.

Matthieu: The idea that we could dispense of the most fundamental quality of consciousness seems quite strange. From the Buddhist perspective, the mind's ability to act on itself, transform itself, recognize its basic nature, and gain freedom from afflictive mental states is crucial and lies at the center of the spiritual path. It is hard to imagine that someone could achieve such a freedom—which is the same as achieving mastery of mind because freedom is to be in charge of one's own mind instead of being the powerless slave of every single thought and emotion that arises—if consciousness were just an irrelevant epiphenomenon. Let alone spiritual life, considering the central place that conscious experience has in our lives, the idea that it would be only a marginal phenomenon that we could easily dispense with seems strange. Anyway, consciousness is a fact. Without it, our subjective world entirely disappears.

Wolf: Yes, but this argument is not easy to counter. If this ability to be aware of oneself is the consequence of neuronal operations that run protocols on the actual state of the brain, then being aware or conscious of something is the consequence and not the cause of neuronal processes. The neuronal processes would fulfill their functions even without you being conscious of them. Consciousness per se would not be able to influence the neuronal processes, only reflect them. I assume you postulate that consciousness actually influences neuronal processes.

Matthieu: Yes, to transform itself. Roger Penrose spoke of the "emperor of the mind," right? In fact, if consciousness is nothing more than an epiphenomenon, then the mind would be a powerless slave. Consciousness would be nothing more than a small red light that lights up at the end of all brain processes and says, "Okay, I am 'on.'" What would be the use of that?

Please remember that, as David Chalmers remarked, all biological functions, including the articulation of a language that allows communication between two organisms, as well as metarepresentation, can be formulated without having to refer to subjective experience. This finding shows that experience is not a particular moment of objective biological functions but something we are aware of prior to any study of these functions. This makes it difficult to establish a causal relation between neuronal processes and consciousness.

If consciousness did not have the capacity to transform itself, know itself, and work to deeply change its contents, then it would really be worthless. Buddhism starts from the other end of the spectrum: pure awareness. Then it investigates how thoughts, emotions, happiness, and suffering arise from this pure awareness. It tries to understand the processes of wisdom and delusion that are related to recognizing or losing recognition of this pure awareness.

This allows one to maintain the recognition that all mental events arise within awareness simply because of many causes and conditions, which do not belong to pure awareness. Pure awareness is unconditioned, in the same way that space is not altered when clouds form—or don't—within it. As I mentioned earlier, pure awareness is a primary fact. You cannot reach experientially to a more fundamental state of consciousness. In this state of pure experience, there is not even a hint of any relation whatsoever with the brain or any other biological process.

Pure awareness is what allows all mental constructs and discursive thoughts, but it is not a construct itself. It leads you to recognize that, thanks to this, you always have the possibility to change the content of your mind because mental states are not intrinsically embedded in pure awareness. Consequently, with training and mindfulness, one can get rid of hatred, craving, and other afflictive emotions.

Wolf: How can we conceive of the possibility that a phenomenon that is the result of neuronal operations, which we experience as a conscious experience, acts on these neuronal processes to change them through a top-down causation process? I would posit that it is the neuronal activity associated with the conscious experience and the ensuing memory traces that impact further neuronal processes. I sense that you postulate that the awareness impacts future neuronal processes.

Matthieu: That's right. Can the basic quality of pure awareness allow for downward causation on mental processes? Can we use it to change our mental landscape? If we consider pure awareness as a primary fact, and there is nothing that goes against this view, there is no reason to deny that mental constructs arising within the space of awareness could act through neuroplasticity. Thus, through the work of interdependent, mutual causation, one may have downward, upward, and same-level causation.[1]

Wolf: I shall have to think about this in greater depth tonight. I am inclined to believe that being conscious or aware is a particular state of the brain that enables the engagement of processing modes that differ substantially from those that do not give rise to awareness. Conscious processing allows for a particularly high degree of information integration. In the workspace of consciousness, signals from various sensory domains can be compared with one another and bound together. This is a prerequisite for the formation of abstract symbolic representations and the likely reason for the close relationship between the conscious processing mode and the linguistic reportability of consciously processed material.

Matthieu: But all these are complex functions—symbolic representations and so on—that do not explain at all the experience of pure awareness, which is the most vivid state of consciousness and is entirely devoid of complexity. But let's call it off for the day.

***

Matthieu: It is a beautiful morning, and we just saw the majestic Himalayan peaks come out of the clouds.

Wolf: Yes. I wish we could just stay in this state of phenomenal awareness rather than having to worry what its function could be.

Yesterday I defended the position that consciousness is an emergent phenomenon of cognitive processes in the brain relying most likely on the generation of metarepresentations, the reapplication of cognitive operations on the results of first-order cognitive processes that mediate primary perception. This iteration and reapplication of cognitive operations might enable the brain to become aware of the results of its own cognitive processes without having any insight into the mechanisms through which these cognitive operations are achieved. We have no recollection of the mechanisms, but we are apparently able to reprocess the results of early cognitive operations until we become aware of being a cognitive system that can have perceptions, feelings, and states. I questioned whether this awareness would also develop if human beings were raised in complete social isolation, without being embedded in a differentiated sociocultural environment, without the possibility of experiencing the effects of one's actions on others, without the possibility of the conceptual framing of one's experiences through language, and without the possibility of developing one's self-model through interactions with peers who have already internalized culture-specific concepts in their own self-model.

Actually, I doubt that meta-awareness would develop without this embedding in a rich sociocultural environment. Rather I believe that meta-awareness is a cultural achievement and that the ability to be aware of being aware results from epigenetic shaping of cognitive brain architectures, from developmental processes shaped by experience. Neuronal architectures underlying basic sensory functions are shaped by experience and interaction with the outer world. The same could be true for neuronal networks supporting the higher cognitive functions required for the development of meta-awareness. The difference would be that in this case the shaping "environment" is the cultural world, with its social realities, traditions, concepts, and beliefs.

However, even if the development of higher cognitive functions is shaped by interactions with a cultural environment, the mental phenomena emerging from these cognitive abilities are still a consequence of neuronal processes in the brain—at least neurobiological evidence suggests this theory to be true. At present, we have no reason to assume that any immaterial or nonsubstrate-dependent states or forces act on material neuronal processes. If there were, then this would be entirely incompatible with the known laws of nature—hence, the resistance of neurobiologists to the idea of top-down causation, to the possibility that an immaterial "consciousness" could influence neuronal processes. However, as I have tried to argue, efficient mechanisms exist through which "immaterial" social realities, collectively shared concepts and beliefs, can influence brain functions and cognition.

As I proposed earlier, the evolution of brains capable of performing cognitive operations that we address as "conscious" permitted modes of social interactions that ultimately catalyzed cultural evolution with all its consequences on the further differentiation of human cognition. One could talk about top-down causation in the sense that the immaterial constructs of cultures, the social realities, influence brain functions. In this case, the mechanisms are well established and not in conflict with the laws of nature. The belief systems, norms, and concepts shared by a society influence the self-understanding and actions of its members. They act directly on the members' brains through the exchange of social signals. Moreover, they imprint the brains of the next generation through education and epigenetic shaping and thus also have long-term effects on brain functions.

Matthieu: Top-down causation is a problem only in the context of a dualistic stance positing material matter, which is supposed to exist as solid reality—a notion that is challenged by both Buddhism and quantum theory—and a supposedly "immaterial" consciousness, which would be a bizarre, indefinable phenomenon devoid of any status. For Buddhism, both matter and consciousness belong to the world of forms. Both exist inasmuch as they are manifest, but they are also devoid of intrinsic, solid reality. Hence the statement, "Void is form, form is void." Consciousness cannot be reduced to gross matter because consciousness is a prerequisite to conceive of matter and make any description of it.

Michel Bitbol once explained to me that the challenge postulated by Chalmers is that one can propose an objectivist, neurobiological explanation of all processes of cognition, perception, and action without ever referring to the fact that these can be accompanied by consciousness and considered as lived experience. Hence, one cannot argue that consciousness is nothing more than a particular state of the brain. What is associated with the brain and what neuroscience can detect are the cognitive functions that consciousness can perform, such as memorization, conceptualization, and verbal expression of one's experiences.

Wolf: The state of consciousness is clearly a special brain state that differs from the states that mediate cognition, perception, and action, processes that do not have to be accompanied by conscious experience. The central question is, do conscious states because of their special character act on brain processes in ways that differ from those through which subconscious processes influence future neuronal states?

Matthieu: You mean, can there be downward causation?

Wolf: I tried to give an explanation for top-down causation that enlarged the scope by including the realm of social realities. Within a brain-centered framework, the question is whether conscious brain processes provide options for the management of information that nonconscious processes lack. We know that subconsciously processed contents have effects on future brain processes. If one performs a well-practiced skill, such as skiing or playing tennis, one inevitably makes small errors that one corrects, without being aware of having made an error or having corrected it. Still, this error correction will change the motor program slightly; there will be a trace left in procedural memory so that in the future, if the same situation occurs, one will perform better. I think the same explanation would account for the possibility that a conscious state influences future brain processes. However, there must be something special about the conscious state with respect to the quality or nature of encoded information acting on future brain states. We should explore what these special functions are and what their adaptive value could be. My proposal was that conscious processing supports a higher degree of integration of diverse sources of information than nonconscious processing.

Matthieu: You have presented many interesting descriptions of what processes could reasonably account for the mind being aware of itself. However, can these descriptions answer the argument put forward by the Dalai Lama, as well as Husserl and phenomenologists like our friend Michel Bitbol, who says that consciousness precedes anything we could ever say about it and precedes any possible perception or interpretation of the phenomenal world? We cannot step out of consciousness to examine it as if it were merely one aspect of our world. Even if one of those monumental projects succeeds in mapping and describing every single neuron and connection in the brain, it will not tell you much about pure experience.

It's quite clear that the vast majority of neuroscientists have the conviction that, one day, enough will be known about the brain so that all aspects of consciousness will be clearly understood from a neurological perspective. But this conviction stems from the fact that they only consider solving what Chalmers and others call the "easy problem," which is to explain functions attributed to consciousness by means of neurophysical processes. However, they do not touch the "hard problem," which is to explain the lived experience we have when we see the color blue or feel love or hatred. In short, they explain the modalities of consciousness but not consciousness itself. Other Western philosophers of consciousness came to the same conclusion. Cohen and Dennett, for instance, write, "Far from being a formidable obstacle to science, [the hard problem] achieves its apparent hardness by being systematically outside of science, not only today's science but any science of the future ... because it is the products of cognitive functions (i.e., verbal report, button pressing, etc.) that allow consciousness to be empirically studied at all."[2] The Dalai Lama concurs when he says, "We risk objectivizing what is essentially an internal set of experiences and excluding the necessary presence of the experiencer. We cannot remove ourselves from the equation. No scientific description of the neural mechanisms of color discrimination can make one understand what it feels like to perceive, say, the color red."[3]

Buddhism offers a few arguments, which should be investigated, to point out that the qualia of consciousness might not be reducible to pure brain function. As I mentioned earlier, these arguments cannot be dismissed as mere Cartesian dualism.

Wolf: I have a problem with these radical phenomenalist stances. Of course, one is free to take such a position, but one should also be aware that it cannot be backed up by any evidence. It is based on arguments derived from deliberations that follow logically from unproven dogma (i.e., that consciousness precedes everything else). I hope I succeeded in showing how the "hard problem" of consciousness can be softened within a naturalistic description system if we consider that mental phenomena owe their existence to the coevolution of cognitive agents and culture. Neuroscience can elucidate the mechanisms that underlie the cognitive functions that allowed humans to initiate cultural evolution; the humanities investigate the dynamics of cultural systems and the emergence of new concepts, norms, and self-models; and the still young branches of the neurosciences, the so-called social neurosciences, begin to investigate how the embedding of human brains in a sociocultural context acts on the development of brains and their functional differentiation.

But let me return to the question of whether conscious processing, the ability to be aware of one's past and present acts, feelings, and experiences, adds something that nonconscious processing lacks. I think the emergence of consciousness has an advantage with respect to the cognitive abilities of brains, an advantage that I related to the ability of abstraction, of symbolic coding, an advantage that becomes particularly important when it comes to forming complex societies. This suggests that there may have been a coevolution of mechanisms supporting the emergence of conscious behavior on the one side and the formation of societies on the other, the two developments mutually supporting each other.

It is commonly held that a conscious state is a unifying state in which a certain number of contents are bound together to form a unitary experience. What appears in consciousness is coherent. That may be related in an interesting way to brain states because, as far as we know, conscious states are associated with a high degree of coordination of distributed brain processes. In electrographic recordings, a conscious state is associated with a high degree of coherence and synchrony. Let's pursue the hypothesis that the contents which enter consciousness get bound together in a peculiar and unique way, which may not be the case for subconscious contents.

Matthieu: But by doing so, you take it on yourself to define consciousness as the function that synthetizes various cognitions. But this definition does not say anything about pure consciousness devoid of content. Then you say that the "contents that enter consciousness get bound together." On the one hand, you say that consciousness is the result of synthesizing various cognitions, and, on the other hand, that various contents, which are the same as cognitions, "enter" consciousness.

Wolf: I am aware that even experimental scientific approaches are ultimately based on preconceptions that cannot be proven. Still, the scientific approach has explanatory power. It produces insight that goes beyond what one could have derived from mere deliberation or contemplation. As long as we remain aware that we are the observers, instrument builders, hypothesis formulators, interpreters, and rule formalizers, as long as we are ready to admit that our conclusions are valid only within defined borders, and as long as no conflicting evidence exists, we should be fine. Given that the cognitive apparatus that we use for scientific inquiry is the product of an evolutionary process and has been optimized to ensure survival in the world to which it has been adapted rather than to assess "objective truth"—if there is such a thing—we must assume that our cognitive abilities are limited and probably highly idiosyncratic.

Still, if we discover through scientific inquiry certain regularities that can be cast in rules, and if these rules make predictions that can be verified within the set description system, if the artifacts designed on the basis of these rules function as predicted, then we have gained at least some insight. We should modestly take it from there and see how far we get on this empirical and verifiable path, knowing that there are boundaries to what we can know and that we do not know how far these boundaries can eventually be pushed—boundaries beyond which the realm of metaphysics and beliefs unfolds.

I hope this clarifies my position and defines what I mean when I talk about evidence. After this necessary epistemic excursion, I wish to come back to the dichotomy between conscious and nonconscious processing. When we operate in the unconscious mode, we can orient toward stimuli irrespective of the modality, visual, auditory, or tactile, and readily analyze their behavioral relevance. However, little integration across modalities exists. Stimuli are processed more independently than if they had entered consciousness. Had they been processed consciously, they would have been bound together and formed a unified coherent percept.

To achieve this integration, the signals from various sensory systems have to be encoded at a sufficiently abstract level, in a sufficiently homogenous format, so they can be bound together. The evolutionary addition of new cortical areas may have provided the substrate necessary for the integration of signals processed by segregated sensory systems. By virtue of integrating and comparing signals from different modalities, it becomes possible to arrive at more abstract, symbolic descriptions of objects and properties. Cross-modal integration also allows one to discover that objects have invariant properties even though they appear differently in different modalities.

The addition of these novel association areas of the neocortex that permit cross-modal integration prepared the ground not only for more unified, polymodal representation of cognitive contents but also for symbolic coding, which in turn is a prerequisite for the development of a symbolic language system and abstract reasoning. Thus, one might consider consciousness, or the state of conscious processing, as a state where distributed computational results can be bound together into a coherent whole, establishing multiple, simultaneous relations among the various distributed items. This obviously allows for a more abstract, symbolic, and comprehensive description of conditions. The adaptive or fitness value of this advanced processing strategy is obvious.

Moreover, if this unified, condensed, and abstract information can be funneled into a versatile communication system, the evolution of cooperating societies will be greatly facilitated. Contents that are processed consciously have access to the language system, unlike the contents of subconscious processing. If the numerous signals from different sensory modalities have already been bound together into coherent wholes, it is easier to communicate what one has perceived and, because of the reflexive nature of awareness, it is easier to convey information about one's internal state.

Thus, one might consider consciousness as the platform where the distributed computational results, once they have been processed into a common abstract format, can be bound together into a coherent whole and communicated in an economic way. The capacity of this platform is obviously limited, as is the case with all physical information-processing systems. Not all the computational results obtained in the various brain centers at any one moment in time can be represented in consciousness and bound together in a unified way. Thus, one needs attention—to select those results one wants to bind together in a conscious format, in a format that can be communicated through language and stored as explicit "knowledge" in episodic/biographic memory, whence it can later be recalled and again become part of conscious deliberations.

If you give me a report of what is currently in your consciousness, then I can attend to this stream of symbols, decode them, transfer the semantic content to my platform of consciousness, and, if required, store it in my declarative memory. In a sense, I copied a condensed, highly symbolic description of your actual state into my brain and stored it there, and from this moment on, your internal state can act on future processes in my brain. Of course, the same is true for the highly processed and abstract contents that have been unified in my own consciousness and stored in my declarative memory. Each of them will change my brain forever and thereby bias future processes. This is how the special computational abilities that are only realized on the platform of consciousness become integrated in the brain processes of "lower" order. This is how top-down causation is implemented. It relies on the conventional mechanisms of signal transduction and storage.

Matthieu: Buddhism says that the ultimate nature of consciousness is beyond words, symbols, concepts, and descriptions. You may speak of pure awareness devoid of mental constructs, but this is like pointing a finger at the moon and calling it the moon itself; unless you have a direct experience of this pure awareness, these words are empty. What you said is all fine about investigating how mental contents get formed, processed, and integrated and how mental activity relates to the world and others. Nevertheless, I think it doesn't address the nature of the fundamental aspect of consciousness. The mechanics of integrating our experience into concepts, how we build up memory and how we engage in transpersonal communication, can be investigated by both contemplatives and neuroscientists at various levels. An incredibly complex and fascinating body of knowledge has yet to be discovered. But still one needs to go deeper into the most basic phenomenon of being conscious.

Wolf: But is it so mysterious? Imagine you have a platform where you can represent visual contents. You close your eyes so there is no more input to the platform, but there is still a platform where vision can express itself. Couldn't it be the same with the platform of consciousness or even meta-awareness? Because you have a highly evolved brain, you have a workspace of consciousness, a platform, a substrate, or a functional state that permits the binding together of widely distributed, highly abstracted representations. You know that you have this ability because you had conscious experiences previously. If you now succeed in preventing the intrusion of any content onto that platform, then you still have awareness of a workspace into which you can load contents by intention or attention.

Matthieu: However if you are aware of the workspace, it means that this awareness in not "in" the workspace but is more fundamental. So, once more, nothing here can explain the fundamental experiential quality of pure awareness devoid of any conceptual content. Contemplatives who have mastered the capacity to clearly identify this pure awareness or this platform without content describe it as vivid and fully aware, as having a quality of peace. They see that thoughts arise from the space of awareness and dissolve back into it, like waves that surge from and dissolve back into the ocean. The mastery of this process eventually leads to people who have an extraordinary emotional balance, inner strength, inner peace, and freedom. So there must be something quite special in having access to such deep levels of mental processes.

Wolf: I am struggling to see the problem. When we are awake, we are aware of brain states. We are aware of being awake, we are aware of being conscious, ready to engage in all the processes in which an awake brain can engage, ready to direct our attention to internal states or external stimuli. Why then should it be so different to focus one's attention on just this state of awareness? I cannot follow your argument that this state of pure awareness is entirely devoid of content. As you say, it is full of blissful feelings, for which people have found names such as peace, timelessness, unification, unboundedness, and so on.

It is just a particular state of consciousness in which the usual contents are repressed in favor of others. It is an altered state of consciousness that can apparently be trained. Could induction of this state not depend on similar neuronal mechanisms as the induction of the many other states of consciousness that we know exist? Take as examples the altered states of consciousness inducible through hypnosis, auto-suggestion, or rituals. All these practices deliberately manipulate the flow of sensory signals and the entry of contents into consciousness by engaging focused attention.

Let me offer another explanation. Attention apparently has dual functions. One function is the selection of signals from the outer world. The other is to select contents for further joint processing in consciousness. It is well established that one can train attention to stabilize one's interactions with the outer world. Some diseases render it difficult to focus attention and not get distracted. One purpose of education is to train our children to learn to focus their attention, filter sensory signals, and concentrate on tasks.

In much the same way, it should be possible to train the attentional mechanisms that help one select the contents that are to be processed at the conscious level and avoid intrusion of unwanted material. Maybe the practice you advocate is not so much the training of the attentional mechanisms that select stimuli from the outer world but of the mechanisms that regulate access to consciousness. At this stage, I can only speculate because too little is known about the organization of attention mechanisms. Also, to the best of my knowledge, we ignore whether separate attention systems exist for the selection of sensory signals and the gating of access to consciousness. Usually, the two processes are tightly linked. What is attended to tends to be processed consciously, has access to working and episodic memory, and can be verbalized. Another caveat is related to the concept of a workspace. When I say *workspace* or *platform*, we should not think of an anatomical site that can be pinpointed. Rather we should think about a particular dynamic state in which binding functions can be realized that characterize conscious processing. It would be one role of attentional mechanisms to prepare that state. Does that make sense?

Matthieu: Yes, it does. Forgive me for insisting, but we still need to explain the nature of basic awareness.

Wolf: That could just be the attention-dependent preparation of that workspace, the setting of the frame.

Matthieu: We would not say that pure awareness is a form of attention because attention implies being attentive to something, in a dualistic mode of a subject who pays attention to an object. It is more correct to say that within pure awareness, various mental functions can unfold, including attention focused on perceptions or any other mental phenomenon. What you say seems to be a satisfactory explanation of the training aspect of attention. But that may not be enough to explain the whole range of experience, especially the fact that experience always comes first. This fact is inescapable fact.

PUZZLING EXPERIENCES

Matthieu: It would be interesting to consider phenomena that would, if they were valid, make us reconsider the general assumption that consciousness is entirely dependent on the brain. Off the top of my head, I can think of three that merit consideration and for which we certainly need to distinguish illusions from reality, fact from rumor: people having access to the content of someone else's thoughts; people describing memories of past lives; and people having near-death experiences or reporting on details of their surroundings at the time that they were apparently unconscious, with a flat EEG, which suggests an absence of electrical activity in the main parts of the brain. Because these phenomena are often cited as factual evidence that consciousness is not limited to our physical body, we need to at least consider what would constitute a valid criterion for such evidence. However, as the work of Steven Laureys has shown, many kinds of comatose states exist, and in some of them, it could be shown that the person can actually be aware of his surroundings.[4]

Wolf: This epistemic issue is important indeed. If any of these reports on parapsychological phenomena were valid and resisted trivial explanations such as misperception, false memory, or coincidence, then we would face a major problem because these phenomena cannot be accounted for by any of the known neuronal mechanisms—and even worse, they would violate some of the fundamental laws on which our natural sciences are based. One major problem with all these phenomena is their lack of reproducibility. They cannot be generated intentionally and hence cannot be investigated experimentally. One could of course again argue that they belong to a class of phenomena whose constitutive property is irreproducibility, that they are singularities of a dynamic that never repeats itself. In that case, one can't study them with the scientific tools at hand.

I shall tell you a story of my own life that still intrigues me. My kids were around eight years old and were invited to a carnival party far out at the other end of town where I had never been before. They were driven there by parents of one of their classmates, and I was supposed to get them in the evening. I left the lab and drove through a snowstorm to the address given to me, which took about an hour.

I arrived at the house, but it was dark and empty. Maybe I remembered that these people had moved, but even this I can't recall consciously. So the situation was highly unpleasant. I had a wrong address, my wife was not at home, and cell phones were not used at that time. I would have to drive back and wait until the kids called to give me a new address. This meant driving one hour back, driving one hour again to get them, and then driving another hour home. I was furious. So what did I do? I continued driving farther out of the town, taking a right, taking a left, stopping at red lights, going somewhere, in an altered state of consciousness. I ended up on a dead-end street, had to make a U-turn, drove back a few hundred meters, and then for some reason felt the urge to park on the right side of the street, which was cumbersome because of the snow. There was a multistory building on the opposite side of the street. I crossed the road and read down the doorbell name plates—don't ask me why I chose this house—and while I was reading, I saw movement in my peripheral vision.

I turned around and through the glass door saw one of my daughters coming up from the cellar, where they had the party, to take the elevator to go up to the apartment and get her coat. I saw her, and I banged at the glass. She opened the door and said, "You are just in time. We are about to finish; Tania will come up very soon." Then I told them the story, and they were not surprised at all! They said, "You are our father, of course you know where we are." Was there some unconscious knowledge, like me having seen the town map at some stage, having heard that the family had moved, subconsciously remembering the name of the road, remembering that they were in a high-rise now and no longer in an individual house? Could it be that I had all that data stored subconsciously and that my subconscious relied on the heuristics that, rather than driving for another three hours, I might do better to wander around a bit, chances being not too low that I might actually find my daughters? But my perception was that I turned at random and did not know why I had engaged in such an irrational search process.

Matthieu: You were not even looking at the street names?

Wolf: No. I was furious. I was just driving around, and I had no explicit memory of the street name of the new address. The interpretation that I was able in that altered state of consciousness to recollect a wealth of data from my subconscious stores and use them for the search process would be compatible with what we know about brain mechanisms. The explanation of my daughters would not. If their interpretation were valid, we would have to worry about our views on the brain and nature in general and admit that we are missing something essential. For the time being, however, and despite such experiences and reports, we don't have a strong argument to change the direction of our research because we wouldn't know what to look for.

Matthieu: You certainly must have felt odd about it.

Wolf: Very odd, yes.

Matthieu: But as a respectable scientist, if you put too much emphasis on this experience, people might start saying, "Hey, look, Wolf Singer is also one of those lunatics who believes in weird phenomena."

Wolf: For sure, if I started a research program on these topics. But let me repeat, there may well be a simple explanation. The storage capacity of our brain is enormous, and we constantly rely on information that we are not aware of in our attempts to remain unharmed through our daily lives. In doing so, we rely on heuristics that are highly efficient but differ from strategies that we consider rational. If these heuristics fail, we tend to take this for granted. However, if they succeed, the solutions are often experienced as miracles.

Matthieu: Let me tell you a personal story. I have told this story before because I think it is a perfect example of what is possible. While I was living in a small hermitage near my first teacher, Kangyur Rinpoche, in Darjeeling, I remembered one day how when I was a teenager, I'd killed a few animals. I used to go fishing until I realized one day when I was 13 that I was inflicting terrible suffering on the fish and depriving them of their lives. I have never been a hunter, and I was actually strongly against hunting, but my uncle had many rifles, and once I thought it would be fun to shoot at a rat because they were devastating his grounds in Brittany. The rat jumped and disappeared under water—I don't know if I killed it, I hope not, but maybe I did.

Pondering this, I thought, "How could I ever do that?" It was such a senseless act with a total lack of consideration for another sentient being's life. That rat might be a mother with some pups. ... I felt a very deep regret over having possibly taken a life for no other reason than the rat was eating the grass of my uncle's lawn.

Remembering all of this, I felt a strong urge to tell my teacher about it and offer him a confession. I came down from my hermitage and went to the monastery to see Kangyur Rinpoche. At that time, I didn't speak much Tibetan, but his elder son, who is also one of my teachers, was there, and he speaks fluent English. While I was offering three prostrations to Kangyur Rinpoche, I saw him laughing and saying something to his son. When I approached to get his blessing and tell my story, before I could open my mouth, his son told me, "Rinpoche is asking, 'How many animals did you kill in your life?'"

Strangely enough, there was nothing strange about all this. It seemed natural. I then told Kangyur Rinpoche that indeed I had killed a rat and many fish. He just laughed again, as if it were a good joke. There was no need of further explanation.

Once, I told my neuroscientist friend Jonathan Cohen about this. He replied, "You have millions of things happening in your life; every moment there is something happening. Out of these millions of events, on some rare occasion, two seemingly unrelated things seem to match perfectly, like winning the lottery. This makes a strong impression in your mind and you conclude that these two events were related in some kind of mysterious way. But this is just an ad hoc explanation of purely random events."

Wolf: Even the very, very improbable is still probable, and when it happens we often assign enormous value to it.

Matthieu: I must say that such things happened a lot when I lived near my teachers—it was like winning the lottery every other month!

Wolf: Mothers during wartime say, "I dreamed my son was killed, and two days later I got the message." People say mothers of soldiers are always anxious; they probably have that dream every night, and if nothing happens, they just forget about it. The usual explanation is that these "visions" are a posteriori interpretations of mere coincidences.

Matthieu: One may of course argue that many seemingly odd coincidences occur every day but are purely incidental. Every single encounter we have every day in the street, in a train, or anywhere has a probability of one in many millions to occur, but we don't pay much attention to it because it doesn't mean much to us. However, when we happen to sit in a train next to someone we know, someone who normally does not use that train line, we are amazed that such a coincidence happened. There is really no difference between the two. Both situations were equally likely, or unlikely, to happen, but the second one has some meaning to us, and we therefore pay special attention to it.

I can give you two striking examples of this. One day I was walking in the street in Paris to go see my publisher and then go to a literary TV program about a new book of mine that had just come out. Suddenly a taxi stopped, and someone came out of it holding a letter. He said, "I don't know you, but I read your book and was just about to go post this letter to you. Here it is." Okay, fine—just a rare event that happened to make sense to both of us. The same evening, after the TV program, we all went to a restaurant to have dinner. Later I took a taxi home with a friend who was going in the same direction. As we talked about the TV program, the driver turns to us and says, "Two hours ago, I picked up a woman who had been at the TV station to be in the audience." When I asked him where he dropped her off, he gave my sister's address. The driver told me that there were 14,000 cabs in Paris.

I witnessed two quite striking coincidences within a matter of hours. But frankly, I see nothing special about this. It is the same as winning the lottery twice in one day. It does not happen often, but it is nothing other than a question of probabilities. It makes perfect sense even though it is quite amusing. There was plenty of sound reason for this man to write a letter to me and go and post it, for me to walk in the street, for my sister to take a taxi to go home, and for me to do the same two hours later.

In the case of my teacher's question, I think that it is of a different nature. There was absolutely no reason for him to ask me, completely out of the blue, whether I had killed animals in my life. First of all, over many years, Kangyur Rinpoche never, ever asked me about particular details related to my childhood or my life in France. All I had ever told him, when I first met him, is that I had done some studies in Western science, and that I had both my parents, an uncle, and a sister dear to me. That's it.

For years, everything he told me was about meditation practice, the life stories of past great masters, or things related to the present moment and daily life. So why in the world would he, for the first and last time in seven years, ask me abruptly about some relatively obscure, distant event of my teenage life? Not only that—which seems quite bizarre in itself—but ask me right at the moment when I was about to tell him about this event that had sprung into my mind only a few moments before? For me, there is something more than a simply probabilistic interpretation, as with the events in Paris, and the simplest and most obvious explanation was that he read my mind. For me, there was no question about this.

This was not even an isolated case. I could tell you of four or five similar stories that I witnessed involving my second main spiritual master, Dilgo Khyentse Rinpoche, and some of his disciples.

In the Tibetan tradition, this is considered to be a secondary effect of a refined level of meditation. Such capacities are not found in ordinary, untrained people. Spiritual teachers will never boast about such capacities—as some "psychics" might do—or even openly acknowledge them. It simply happens from time to time, maybe when the moment is appropriate as a way to strengthen the disciple's confidence in the spiritual practice. It is always a subtle hint, never an ostentatious display.

Wolf: Experiments were performed in the United States during the Cold War at Stanford University. They were trying to figure out how to communicate with submerged submarines, and the idea was to exploit telepathy. Serious physics laboratories engaged in experiments. A target person was sent to one of five defined places, and a medium seated in a shielded room, a Faraday cage—an enclosure formed by conductive material used to block electric fields—had to give a verbal and a pictorial description of what he thought the target person was seeing in that moment. Then this material was given to a group of people who had been familiarized with these five places but didn't know anything about the experiment. They were asked to find out which place most resembled the verbal report and the drawings made by the medium.

The statistics were apparently without flaws, and the hit rate was fantastic. There was a highly significant correlation between the judgment of the impartial observers who rated the pictures and the respective place where the target person had actually been sent. Two of these papers were published in *Nature* and one or two were in *IEEE*—a respectable journal of the engineering community. These publications had no follow-up; at least I haven't heard anything about them since.

Matthieu: That was when?

Wolf: It must have been in the 1960s. It seems these physicists had practiced good science, with a double-blind design and all. An institute in Freiburg, Germany, run by Professor Bender, also tried to prove these strange phenomena in experimental settings. However, in this case, all attempts failed, and the statistics never became significant.

Matthieu: The question is, what do you do with this? It goes against all cultural belief in the West (at least against the established scientific culture), and so people tend to just disregard these things even though they have not been invalidated.

Wolf: I remember 20 years ago, it was considered unserious to explore the neuronal substrate of consciousness. Now it is an accepted field of research. But I don't see any granting agency giving you a penny if you came up with the idea of studying parapsychological phenomena. The *Nature* paper of the Stanford physicist is probably completely forgotten. I talked to a colleague who knew about these publications one evening over a glass of wine. He showed me reports about the research in respectable journals because I was doubtful. The articles looked legitimate, but the problem is that most investigations on parapsychological phenomena eventually turned out to be flawed. I have not followed up on the fate of the Stanford studies.

REMEMBERING PAST LIVES?

Matthieu: Ian Stevenson, a now deceased professor at the University of Virginia, has studied statements made by hundreds of people who claim to have recollections of past lives. He dismissed most of them as being either inconclusive or bogus. However, in the end, after unpacking all the details and facts, he singled out 20 cases for which the usual explanation did not seem to fit. The details and accuracy of the recollections were hard to explain as anything other than the product of some kind of memory.[5] All of these cases were related to ordinary children. Stevenson was not a religious believer but an anthropologist.

One famous historical case is that of Shanti Devi. I often told the story, but it is quite remarkable. She was born in Delhi, India, in 1926. When she was about four years old, she started saying strange things to her parents. She told them that her real home was in the town of Mathura, where her husband lived. Shanti Devi was an intelligent and nice child, and at first people were amused, but they soon started to worry about her sanity. Everybody at school made fun of her.

Eventually, however, her teacher and headmaster became so intrigued by the studious, serious girl that they went to see her parents to clarify the whole affair. They questioned Shanti Devi for a long time. During this conversation, she continuously used words from the Mathura dialect, which nobody in her family or school spoke. Among many other details, she said that her husband's name was Kedar Nath. Intrigued, the headmaster made inquiries in Mathura, found that there was a merchant whose name was Kedar Nath, and wrote to him.

The astonished merchant replied, confirming that, nine years before, his wife had died 10 days after giving birth to their son. He sent one of his cousins to Delhi. The little girl immediately recognized this man whom she'd never seen, welcomed him warmly, told him that he'd put on weight and she was sad to see him still unmarried, and then asked him all sorts of questions. This cousin, who'd come thinking that he was going to unmask an imposter, was flabbergasted. She then asked him news of her own son.

When he heard all this, Kedar Nath decided at once to go to Delhi with his son, with the intention of passing himself off as his brother. But no sooner had he introduced himself under his false name than Shanti Devi exclaimed, "You're not my *jeth* ("brother-in-law" in the dialect of Mathura); you're my husband, Kedar Nath." Then she rushed into his arms in tears. When the son, who was just slightly older than the little girl, came into the room, she kissed him as a mother does. Shanti Devi asked Kedar Nath if he'd kept the promise that he'd made on her death-bed not to remarry. She then forgave him when he admitted having taken a second wife. Kedar Nath asked Shanti Devi a thousand questions that she answered with quite disconcerting accuracy.

Gandhi himself went to see the little girl and proposed to send her to Mathura, with her parents, three respectable townsmen, and some lawyers, journalists, and businessmen, all of high intellectual repute. When the party arrived at Mathura station, a crowd was waiting for them on the platform. At once, the child astonished everyone by recognizing the members of her "former family." She ran over toward an old man and cried out, "Grandfather!"

The girl led the procession straight to her house. Over the next few days, she recognized dozens of people and places. She met her former parents, who were overwhelmed. Her current parents were extremely worried that she wouldn't stay with them. Despite being torn, she decided to go back to Delhi with them. Thanks to her questions, she'd found out that her husband had kept none of the promises that he'd made to her on her deathbed. He hadn't even offered to Krishna her savings of 150 rupees that she'd hidden under the floorboards for the salvation of her soul. Only Lugdi Devi and her husband knew of this hiding place. Shanti Devi forgave her husband for all his failings, while everyone who heard her admired her more and more. A local commission carried out its investigations scrupulously, cross-checking information and accumulating details. It concluded that Shanti Devi was the reincarnation of Lugdi Devi.

Shanti Devi then lived a modest life and, after studying literature and philosophy, gave herself over to prayer and meditation.[6]

So you see, without being fascinated or obsessed by weird events, like people who spend their lives studying UFOs, I feel it is part of having an open mind to investigate such phenomena.

Wolf: We should keep our minds open. The history of science is full of examples where observations were thought to be incompatible with generally accepted theories. These conflicting observations then led to further explorations that finally forced a modification of theories or led to the discovery of entirely new principles. The observations that we have just discussed cannot be accounted for within the framework of contemporary scientific theories. If one were to make them the object of scientific investigation, one would have to be able to reproduce them under controlled conditions or at least find some evidence for the transmission of information across space and time that follows something other than the known principles. If these mysterious phenomena have the properties of singularities, such as occur in complex nonlinear systems (i.e., if it is one of their characteristics to not be reproducible), then conventional scientific approaches fail.

One could of course argue that other nonreproducible processes such as evolution or the development of the universe are readily accessible to scientific exploration and explanations. Why then should this not be the case for the parapsychological phenomena observed and reported by so many? The answer is that evolution of both the universe and biological evolution can be accounted for within the framework of the known laws of nature, whereas the known principles of information transmission across space and time fall short of explaining any of these phenomena, and at present, we do not know where and what we should explore with the tools that we have at hand. We have learned from the history of scientific breakthroughs that it is most often futile to attempt to solve a problem head on if one does not know where to begin. The more promising strategy is to simply continue to search where the light is (i.e., where testable hypotheses can be formulated). If, in the pursuit of this strategy, novel principles were discovered that could give a hint as to the processes underlying some of these parapsychological phenomena, the time would have come to subject them to scientific investigation. At the moment, we have not reached that point.

WHAT CAN BE LEARNED FROM NEAR-DEATH EXPERIENCES?

Matthieu: The third kind of testimony that would be interesting to investigate is the near-death experience (NDE). It seems that many things experienced by people who had an NDE (the experience of bliss, seeing lights at the end of tunnels, experiencing oneself as floating above one's own body) could be easily explained by the sudden rush of neurotransmitters that occur in the brain when one is at the edge of dying. A few documented reports, including a review published in the reputable medical journal *The Lancet*, also describe people seemingly recollecting some events that happened in their hospital room while they were in a coma with a flat EEG. Pim van Lommel, the author of this review of 354 cases of patients who had a cardiac arrest,[7] reported that one of the patients clearly remembered a nurse taking away his denture on a tray while his brain was apparently inactive. When he woke up from his coma, the nurse was not on duty. But when she turned up a few days later, the patient said to her, "Hey, where did you take my denture?" She was quite shocked.

Wolf: Numerous accounts, published in serious journals with peer review, include patients who report such experiences after being reanimated following cardiac arrest or other incidents that temporarily abolished measurable brain activity. In most cases, these experiences can be attributed to the massive disturbances of brain functions that precede and follow the comatose state. In general, it is impossible to know when exactly these NDEs occur: during the phase of falling into the coma or during the phase of awakening. The patients' estimates of timing are not reliable. When a brain falls into and comes out of a coma, the sequences of switching off and on the brain's various subsystems do not follow the normal order that characterizes falling asleep and awakening—hence the dissociated experiences and confusion of time and space.

As to van Lommel's report, I have no explanation. A speculative interpretation could be that the patient had seen the nurse before falling into the coma and somehow had some subconscious recollection of somebody taking his denture out of his mouth—which is a quite invasive procedure, especially if low-level oral reflexes are still present—and that subconscious processing established a link between these temporally contingent events. Also, we know from well-controlled experiments performed during anesthesia that patients can have subconscious recollections of events that occurred during surgery, even though the electroencephalogram showed clear evidence of deep anesthesia, a state that closely resembles a coma. The same, by the way, holds for deep sleep. Although the sleeping person is unconscious, the brain is still capable of analyzing sensory signals, and if any are identified as unusual or dangerous, to respond with awakening. As you mentioned, that is what Steven Laureys's and others' work has shown.

Again, however, if there is convincing evidence that subjects experienced and remembered events that actually occurred while their brain was completely inactive—which is not easy to document with electroencephalographic recordings alone—then we have a similar problem as with the parapsychological phenomena. Also relevant in this context is the abundant evidence that the abnormal brain activity which occurs during the aura of an epileptic seizure—the interval preceding a generalized seizure—may cause patients to experience phenomena that clearly resemble those reported in an NDE. Eloquent literary testimonies of this experience are given by Dostoyevsky in his novel *The Idiot*, in which he describes succinctly the altered states of consciousness preceding his seizures. He experienced these states as extremely rewarding, blissful, and of great clarity. All conflicts seemed to be resolved, and he felt himself in perfect harmony with the world, unbounded, endowed with great insight, and living in the absolute present.

Strikingly similar experiences are described by patients who suffer from focal epilepsy in the anterior insula, an area of the neocortex that is strongly connected with areas in the prefrontal cortex, the anterior cingulate cortex, and reward systems.[8] One of the numerous functions attributed to this area is the detection of mismatches between expected or predicted and actual brain states. Because epileptic activity disrupts the functions of the affected regions, one interpretation of the ecstatic and blissful feelings associated with insular seizures could be the transitory lack of error signals. This could give rise to the blissful feeling of clarity, the feeling of having resolved all conflicts, and being in perfect harmony with oneself and the outer world. One patient described this state as being like an uninterrupted sequence of "eureka" moments, the satisfactory happy moments that one experiences after having found a solution to a problem or having obtained a sudden insight that we talked about earlier.

COULD CONSCIOUSNESS BE MADE OF SOMETHING OTHER THAN MATTER?

Wolf: If top-down mental causation were possible, we would have to postulate a process supported by an unknown dimension of nature. This "something" would have to control neuronal processes and shape them so that they materialize in our thoughts, wishes, and emotions, and all traits of our personality. In that case, this process would not be bound to our body and brain, and it could ensure our continuity beyond death.

Matthieu: It would be the vector of the continuation of consciousness.

Wolf: Then the question is, How does this process interact with the sophisticated neuronal networks of my brain so that these execute what this supra-ordinate process has "in mind"? In addition, and complicating explanations of telepathy even further, the networks of different brains differ substantially not only because of genetic diversity but also because they have been raised in different environments and undergone different epigenetic shaping processes. How would it then be possible that you know how to generate the waves or force fields that influence my brain in a distinct manner? None of these questions is addressable within the framework of scientific theories that I am aware of.

Evidence suggests that the electrical fields generated by synchronized neuronal activity are strong enough, even though they are extremely weak, to bias the activity of other neurons in the vicinity. It is still unclear whether this nonsynaptic—we call it ephaptic—modulation of neuronal activity is functionally relevant. During epileptic seizures, the fields generated by the highly synchronous discharges of a large population of neurons are strong enough to actually entrain adjacent neurons even if these are not synaptically connected. So the possibility exists that the fields generated by neuronal populations in turn act on their own activity. However, these effects are local, confined to distances of a fraction of a millimeter, and so far no known "fields" could mediate long-distance effects. One can influence neuronal activity across the skull, but to do so one has to apply strong electrical fields either via electrodes placed on the scalp or by generating extremely strong electromagnetic pulses close to the skull, which induce currents in the underlying brain tissue. But these methods are coarse and can only induce global changes of excitability.

Matthieu: That's indeed interesting, but from a Buddhism perspective, this related to the "gross" aspect of consciousness, not to its fundamental nature.

Wolf: Yet these facts go against the possibility of mental phenomena that are disconnected from and independent of the material substrate of the brain. It really challenges the core of most religions, which postulate the existence of an immaterial mind or soul that is independent of the brain and dwells in its own spiritual, immaterial realm but interacts with the brain.

There is of course another way to conceptualize the idea that we have a dimension in which our "mind" or "soul" persists beyond our own physical existence and influences other brain functions. As discussed previously, through our social and scientific activities, we added a wealth of new realities to the world in which we evolved. Many of these realities are immaterial: They consist of beliefs, concepts, knowledge, agreements, symbols, attributions, value assignments, and so forth. Each of us contributes through participation in culture building to these immaterial realities, and they persist beyond our own physical existence. As they say, "No word is ever lost." Moreover, although immaterial, these realities strongly influence the functioning of others' brains. They create constraints, moral imperatives, and social goals for other members of the society, and they can even impact the functional architecture of the brains of the following generations by experience-dependent epigenetic shaping of brain development.

We discussed the mechanisms through which immaterial social realities can modify brain architectures during development. With respect to genetic heritage, our brains do not differ too much from those of our cave-dwelling ancestors. However, because of our embedding in a much richer and more complex sociocultural environment and a host of effective educational measures, the architecture of our brains is with all likelihood more differentiated. In other words, cultural activities, activities of the "mind," create immaterial realities that in turn act on brains and their architectures. Maybe implicit intuitions of these processes gave rise to concepts and beliefs in an immortal, immaterial dimension of our existence, our soul, the possibility of mental causation and reincarnation.

Matthieu: A close relationship undoubtedly exists between the neural workings of the brain and what Buddhists would call the "gross aspect" of consciousness. This is why the brain's physical condition so profoundly affects this type of consciousness.

That being said, scientific theories are inevitably influenced by the metaphysical viewpoints prevailing in the culture in which they are devised. Most Western scientists and philosophers tend to believe that a solid reality exists behind the veil of appearances. This is why quantum physics, according to which particles are not things but events that can behave either as nonlocalized waves or as localized particles, is deeply puzzling to anyone with a belief in a "solid" world. Physics presents us with a number of worldviews, among which realism is but one of several possibilities. Eastern cultures have less difficulty in calling into question the solid reality of the phenomena world. It is also easier for them to conceive of the existence of the fundamental, primary levels of consciousness—pure awareness or pure experience, as they are millenaries old tradition of investigating the mind through direct introspection.

Our mutual friend Francisco Varela wrote on this subject, "These subtle levels of consciousness appear to Western eyes as a form of dualism and are quickly dismissed. ... It is important to note that these levels of subtle mind are not theoretical; instead, they are delineated rather precisely on the basis of actual experience, and they merit respectful attention by anybody who claims to rely on empirical science."[9]

Francisco once told me that, with regard to the ultimate nature of consciousness, it would be wise to keep an open mind so that we don't overlook all the various modes and explanations that could account for consciousness.

In the spirit of open-mindedness that we have nurtured throughout our friendship and these most enjoyable encounters, it seems therefore fitting that we leave this vital question open to more discoveries, from both the third- and first-person perspectives.

A CONCLUDING NOTE OF GRATITUDE

Matthieu and Wolf: This book is coming to an end. It marks eight years of conversations that we are delighted to share with our readers. Motivated by curiosity and mutual friendship, we explored some essential questions concerning the nature of the human mind. It was our intention to combine our respective expertise and exploit two complementary sources of knowledge: the first-person perspective provided by introspection and contemplative practice, and the third-person perspective taken by the neurosciences. Of course, we were aware from the beginning that we would fall short of reaching any final answers to the profound questions that humanity has been discussing for thousands of years. However, we hope that we have at least arrived at a clearer identification of some commonalities, as well as some persisting gaps of knowledge. We wish to thank our readers for having been with us to this point.

We also wish to thank all those who accompanied us on this journey. Our publishers first: Ulla Unseld Berkewicz, Nicole Lattès, and Guillaume Allary, who gave us the freedom to carry on our conversation at our own pace, meeting off and on and maturing our ideas over eight years. This is rather unusual in a publishing world, where often the idea of a book that did not exist the day before suddenly becomes something that must be completed by an imminent deadline. We are immensely grateful to our publishers for their patience and support.

Our thanks go to all those who gave us hospitality during our various encounters—the Suhrkamp house in Frankfurt, where our formal discussions started; the mountain hermitages of Shechen Pema Ösel Ling, facing the majestic Himalayan ranges in Nepal, where we sojourned twice; and the beautiful jungle resort of Thanyamundra at Koh Sok in Thailand, where we enjoyed the hospitality of Klaus Hebben.

We thank Suhrkamp Verlag for transcribing our conversations that were held in English, which is not our mother tongue.

We are especially grateful to Janna White, who carefully and wisely edited the English manuscript and brought it to its final form. We are also grateful to the MIT Press team for carefully checking this volume and bringing it to publication in a beautiful way.

Many other friends and relatives accompanied us over the years. We thank them all from the heart.

NOTES

Introduction

1. The Mind and Life Institute was founded in 1987, the result of a meeting of three visionary minds: His Holiness the Dalai Lama, Tenzin Gyatso; Adam Engle, lawyer and entrepreneur; and the neuroscientist, Francisco Varela. The objective of the Mind and Life Institute is to encourage interdisciplinary dialogue among Western science, the human sciences, and contemplative traditions. It aims to support and integrate the first-person perspective, arising from the experience of meditation and other contemplative practices, into traditional scientific methodology. This objective's determining influence is seen in several books: *Train Your Mind—Transform Your Brain* by Sharon Begley, *Destructive Emotions: How Can We Overcome Them?* by Daniel Goleman, and *The Dalai Lama at MIT* by Anne Harrington and Arthur Zajonc.

2. These conversations were held in September 2007 in Frankfurt, in December 2007 and February 2014 in Nepal, in November 2010 in Thailand, and on a few other occasions in Hamburg and Paris.

Chapter 1

1. P. A. Van den Hurk, B. H. Janssen, F. Giommi, H. P. Barendregt, and S. C. Gielen, "Mindfulness meditation associated with alterations in bottom-up processing: psychophysiological evidence for reduced reactivity," *International Journal of Psychophysiology* 78, no. 2 (2010): 151–157.

2. B. Röder, W. Teder-Sälejärvi, A. Sterr, F. Rösler, S. A. Hillyard, and H. J. Neville, "Improved auditory spatial tuning in blind humans," *Nature* 400 (1999): 162–166.

3. G. Kempermann, H. G. Kuhn, and F. H. Gage, "More hippocampal neurons in adult mice living in an enriched environment," *Nature* 386 (1997): 493–495.

4. P. S. Eriksson, E. Perfilieva, T. Björk-Eriksson, A. M. Alborn, C. Nordborg, D. A. Peterson, and F. H. Gage, "Neurogenesis in the adult human hippocampus," *Nature Medicine* 4, no. 11 (1998): 1313–1317.

5. H. Eichenbaum, C. Stewart, and R. G. M. Morris, "Hippocampal representation in place learning," *Journal of Neuroscience* 10 (1990): 3531–3542.

6. J. S. Espinosa and M. P. Stryker, "Development and plasticity of the primary visual cortex," *Neuron* 75 (2012): 230–249; W. Singer, "Development and plasticity of cortical processing architectures," *Science* 270 (1995): 758–764.

7. H. Van Praag, A. F. Schinder, B. R. Christie, N. Toni, T. D. Palmer, and F. H. Gage, "Functional neurogenesis in the adult hippocampus," *Nature* 415 (2002): 1030–1034.

8. E. Luders, K. Clark, K. L. Narr, and A. W. Toga, "Enhanced brain connectivity in long-term meditation practitioners," *NeuroImage* 57, no. 4 (2011): 1308–1316.

9. S. W. Lazar, C. E. Kerr, R. H. Wasserman, J. R. Gray, D. N. Greve, M. T. Treadway, et al., "Meditation experience is associated with increased cortical thickness," *NeuroReport* 16, no. 17 (2005): 1893; B. K. Hölzel, J. Carmody, M. Vangel, C. Congleton, S. M. Yerramsetti, T. Gard, and S. W. Lazar, "Mindfulness practice leads to increases in regional brain gray matter density," *Psychiatry Research: Neuroimaging* 191, no. 1 (2011): 36–43.

10. P. Fries, "A mechanism for cognitive dynamics: neuronal communication through neuronal coherence," *Trends in Cognitive Science* 9, no. 10 (2005): 474–480.

11. A. Lutz, H. A. Slagter, N. B. Rawlings, A. D. Francis, L. L. Greischar, and R. J. Davidson, "Mental training enhances attentional stability: neural and behavioral evidence," *Journal of Neuroscience* 29, no. 42 (2009): 13418–13427; K. A. MacLean, E. Ferrer, S. R. Aichele, D. A. Bridwell, A. P. Zanesco, T. L. Jacobs, et al., "Intensive meditation training improves perceptual discrimination and sustained attention," *Psychological Science* 21, no. 6 (2010): 829–839.

12. J. A. Brefczynski-Lewis, A. Lutz, H. S. Schaefer, D. B. Levinson, and R. J. Davidson, "Neural correlates of attentional expertise in long-term meditation practitioners," *Proceedings of the National Academy of Sciences* 104, no. 27 (2007): 11483–11488.

13. M. Ricard, *Happiness: A Guide to Developing Life's Most Important Skill* (New York: Little, Brown and Company, 2006).

14. Mingyur Rinpoche, *The Joy of Living: Unlocking the Secret and Science of Happiness* (New York: Three Rivers Press, 2007).

15. A. P. Jha, J. Krompinger, and M. J. Baime, "Mindfulness training modifies subsystems of attention," *Cognitive, Affective, & Behavioral Neuroscience* 7, no. 2 (2007): 109–119.

16. A. Lutz, L. L. Greischar, N. B. Rawlings, M. Ricard, and R. J. Davidson, "Long-term meditators self-induce high-amplitude gamma synchrony during mental practice," *Proceedings of the National Academy of Sciences* 101, no. 46 (2004): 16369–16373.

17. P. Fries, "Neuronal gamma-band synchronization as a fundamental process in cortical computation," *Annual Review of Neuroscience* 32 (2009): 209–224; A. Lutz, H. A. Slagter, J. D. Dunne, and R. J. Davidson, "Attention regulation and monitoring in meditation," *Trends in Cognitive Science* 12 (2008): 163–169.

18. P. R. Roelfsema, A. K. Engel, P. König, and W. Singer, "Visuomotor integration is associated with zero time-lag synchronization among cortical areas," *Nature* 385 (1997): 157–161.

19. P. Fries, "A mechanism for cognitive dynamics: neuronal communication through neuronal coherence," *Trends in Cognitive Sciences* 9, no. 10 (2005): 474–480; W. Singer, "Neuronal synchrony: a versatile code for the definition of relations?," *Neuron* 24 (1999): 49–65; P. Fries, J. H. Reynolds, A. E. Rorie, and R. Desimone, "Modulation of oscillatory neuronal synchronization by selective visual attention," *Science* 291 (2001): 1560–1563; B. Lima, W. Singer, and S. Neuenschwander, "Gamma responses correlate with temporal expectation in monkey primary visual cortex," *Journal of Neuroscience* 31, no. 44 (2011): 15919–15931.

20. P. Fries, P. R. Roelfsema, A. K. Engel, P. König, and W. Singer, "Synchronization of oscillatory responses in visual cortex correlates with perception in interocular rivalry," *Proceedings of the National Academy Sciences* 94 (1997): 12699–12704; P. Fries, J. H. Schröder, P. R. Roelfsema, W. Singer, and A. K.

Engel, "Oscillatory neuronal synchronization in primary visual cortex as a correlate of stimulus selection," *Journal of Neuroscience* 22, no. 9 (2002): 3739–3754.

21. O. L. Carter, D. E. Prestl, C. Callistemon, Y. Ungerer, G. B. Liu, and J. D. Pettigrew, "Meditation alters perceptual rivalry in Tibetan Buddhist monks," *Current Biology* 15, no. 11 (2005): R412–R413.

22. L. Melloni, C. Molina, M. Pena, D. Torres, W. Singer, and E. Rodriguez, "Synchronization of neural activity across cortical areas correlates with conscious perception," *Journal of Neuroscience* 27, no. 11 (2007): 2858–2865.

23. F. Varela, J. P. Lachaux, E. Rodriguez, and J. Martinerie, "The brainweb: phase synchronization and large-scale integration," *Nature Review Neuroscience* 2 (2001): 229–239.

24. S. W. Lazar, C. E. Kerr, R. H. Wasserman, J. R. Gray, D. N. Greve, M. T. Treadway, et al., "Meditation experience is associated with increased cortical thickness," *NeuroReport* 16, no. 17 (2005): 1893–1897.

25. J. Boyke, J. Driemeyer, C. Gaser, C. Buchel, and A. May, "Training-induced brain structure changes in the elderly," *Journal of Neuroscience* 28 (2008): 7031–7035; A. Karni, G. Meyer, P. Jezzard, M. M. Adams, R. Turner, and L. G. Ungerleider, "Functional MRI evidence for adult motor cortex plasticity during motor skill learning," *Nature* 377 (1995): 155–158.

26. P. E. Dux and R. Marois, "The attentional blink: a review of data and theory," *Attention, Perception, & Psychophysics* 71 (2009): 1683–1700; N. Georgiou-Karistianis, J. Tang, Y. Vardy, D. Sheppard, N. Evans, M. Wilson, et al., "Progressive age-related changes in the attentional blink paradigm," *Aging, Neuropsychology, & Cognition* 14, no. 3 (2007): 213–226; H. A. Slagter, A. Lutz, L. L. Greischar, A. D. Francis, S. Nieuwenhuis, J. M. Davis, and R. J. Davidson, "Mental training affects distribution of limited brain resources," *PLoS Biology* 5, no. 6, e138 (2007): 131–138; S. Van Leeuwen, N. G. Müller, and L. Melloni, "Age effects of attentional blink performance in meditation," *Consciousness and Cognition* 18 (2009): 593–599. The study in which Matthieu participated was done at Anne Treisman's lab at Princeton University by Karla Evans. Unpublished.

27. H. A. Slagter, A. Lutz, L. L. Greischar, A. D. Francis, S. Nieuwenhuis, J. M. Davis, & R. J. Davidson, "Mental training affects distribution of limited brain resources," *PLoS Biology* 5, no. 6 (2007): 138.

28. S. Van Leeuwen, N. G. Müller, & L. Melloni, "Age effects of attentional blink performance in meditation," *Consciousness and Cognition* 18 (2009): 593–599.

29. For a detailed description of this episode, see chapter 1 of Daniel Goleman's *Destructive Emotions: How Can We Overcome Them?* (New York: Bantam Books, 2003).

30. R. W. Levenson, P. Ekman, and M. Ricard, "Meditation and the startle response: a case study," *Emotion* 12, no. 3 (2012): 650–658.

31. G. Wang, B. Grone, D. Colas, L. Appelbaum, and P. Mourrain, "Synaptic plasticity in sleep: learning, homeostasis and disease," *Trends in Neuroscience* 34, no. 9 (2011): 452–463.

32. W. E. Skaggs and B. L. McNaughton, "Replay of neuronal firing sequences in rat hippocampus during sleep following spatial experience," *Science* 271 (1996): 1870–1873.

33. That is, in the deepest phase of sleep, not during the phase of "paradoxical sleep" (REM) that corresponds to dreams. F. Ferrarelli et al., "Experienced mindfulness meditators exhibit higher parietal-occipital EEG gamma activity during NREM sleep," *PloS One* 8, no. 8 (2013): e73417.

34. A. Lutz, H. A. Slagter, N. B. Rawlings, A. D. Francis, L. L. Grieschar, & R. J. Davidson, "Mental training enhances attentional stability: neural and behavioral evidence," *Journal of Neuroscience* 29, no. 42 (2009): 13418–13427.

35. A. Lutz, L. L. Greischar, D. M. Perlman, and R. J. Davidson, "BOLD signal in insula is differentially related to cardiac function during compassion meditation in experts vs. novices," *NeuroImage* 47, no. 3 (2009): 1038–1046; A. Lutz, J. Brefczynski-Lewis, T. Johnstone, and R. J. Davidson, "Regulation of the neural circuitry of emotion by compassion meditation: effects of meditative expertise," *PLoS One* 3, no. 3(2008): e1897.

36. Other studies suggest that lesions in the amygdala disturb the emotional aspect of empathy without affecting its cognitive aspect. See R. Hurlemann, H. Walter, A. K. Rehme, et al., "Human amygdala reactivity is diminished by the b-noradrenergic antagonist propranolol," *Psychological Medicine* 40 (2010): 1839–1848.

37. O. M. Klimecki, S. Leiberg, M. Ricard, and T. Singer, "Differential pattern of functional brain plasticity after compassion and empathy training," *Social, Cognitive, and Affective Neuroscience* 9, no. 6 (2014): 873–879.

38. B. L. Fredrickson, M. A. Cohn, K. A. Coffey, J. Pek, and S. M. Finkel, "Open hearts build lives: positive emotions, induced through loving-kindness meditation, build consequential personal resources," *Journal of Personality and Social Psychology* 95, no. 5 (2008): 1045.

39. M. Botvinick, L. E. Nystrom, K. Fissell, C. S. Carter, and J. D. Cohen, "Conflict monitoring versus selection-for-action in anterior cingulate cortex," *Nature* 402 (1999): 179–181.

40. Tania Singer is director at the Max Planck Institute for Human Cognitive and Brain Sciences in Leipzig. She is well known for her research on empathy and compassion. Matthieu has collaborated with her extensively over the years.

41. S. B. Kaufman & C. Gregoire, *Wired to Create: Unraveling the Mysteries of the Creative Mind* (New York, NY: Tarcher Perigee, 2015).

42. D. Tammet, *Born on a Blue Day, Inside the Extraordinary Mind of an Autistic Savant* (New York: Free Press, 2007).

43. J. Biederlack, M. Castelo-Branco, S. Neuenschwander, D. W. Wheeler, W. Singer, and D. Nikolic, "Brightness induction: rate enhancement and neuronal synchronization as complementary codes," *Neuron* 52 (2006): 1073–1083.

44. P. Condon, G. Desbordes, W. Miller, D. DeSteno, M. G. Hospital, and D. DeSteno, "Meditation increases compassionate responses to suffering," *Psychological Science* 24, no. 10 (2013, October): 2125–2127.

Chapter 2

1. D. Kahneman, *Thinking, Fast and Slow* (New York: Farrar, Strauss and Giroux, 2011). See also: D. Kahneman, P. Slovic, and A. Tversky, eds., *Judgment under Uncertainty: Heuristics and Biases* (Cambridge, UK: Cambridge University Press, 1982); and D. Kahneman and A. Tversky, "Prospect theory: An analysis of decision under risk," *Econometrica* 47, no. 2 (1979): 263–291.

2. B. Fredrickson, *Love 2.0: How Our Supreme Emotion Affects Everything We Feel, Think, Do, and Become* (New York: Hudson Street Press, 2013).

3. See also the article on this subject by A.T. Beck, "Buddhism and Cognitive Therapy," *Cognitive Therapy Today: The Beck Institute Newsletter* (2005).

Chapter 3

1. Quoted in J. W. Pettit, *The Beacon of Certainty* (Boston: Wisdom Publications, 1999), 365.

2. P. Kaliman, M. J. Álvarez-López, M. Cosín-Tomás, M. A. Rosenkranz, A. Lutz, and R. J. Davidson, "Rapid changes in histone deacetylases and inflammatory gene expression in expert meditators," *Psychoneuroendocrinology* 40 (2014): 97–107.

3. R. Boyd and P. J. Richerson, "A simple dual inheritance model of the conflict between social and biological evolution," *Zygon* 11, no. 3 (1976): 254–262. See also their seminal work: *Not by Genes Alone: How Culture Transformed Human Evolution* (Chicago: University of Chicago Press, 2004).

4. R. Boyd and P. J. Richerson, *Not by Genes Alone: How Culture Transformed Human Evolution* (Chicago: University of Chicago Press, 2004), 5.

5. Ibid., x.

6. T. Nagel, "What is it like to be a bat?", *The Philosophical Review* 83, no. 4 (1974): 435–450.

7. "A reality completely independent from the mind that thinks, sees or feels it is impossible. Even if it did exist, such an exterior world would be forever inaccessible to us." Henri Poincaré, *La Valeur de la science* (The Value of Science) (Paris: Flammarion, 1990).

8. A double-blind study is a study in which half of the patients receive a drug and the other half receive a placebo, but neither the person who delivers the medicine nor the patient knows which of these two is being given.

9. See B. Alan Wallace, *The Taboo of Subjectivity: Toward a New Science of Consciousness* (New York: Oxford University Press, 2004).

10. M. Ricard, *Why Meditate?* (New York: Hay House, 2010).

11. M. Ricard, *Altruism: The Power of Compassion to Change Yourself and the World* (New York, NY: Little, Brown and Company, 2015).

12. R. Tagore, *Stray Birds* (New York: The Macmillan Company, 1916), LXXV.

Chapter 4

1. T. Metzinger, *Being No One: The Self-Model Theory of Subjectivity* (Cambridge: The MIT Press, 2004). See also T. Metzinger, *The Ego Tunnel: The Science of the Mind and the Myth of the Self* (New York: Basic Books, 2009).

2. P. Gilbert and C. Irons, "Focused therapies and compassionate mind training for shame and selfattacking," in *Compassion: Conceptualisations, Research and Use in Psychotherapy*, ed. P. Gilbert (New York: Routledge, 2005), 263–325; K. Neff, *Self-Compassion: Stop Beating Yourself Up and Leave Insecurity Behind* (New York: William Morrow, 2011).

3. W. K. Campbell, J. K. Bosson, T. W. Goheen, C. E. Lakey, and M. H. Kernis, "Do narcissists dislike themselves 'deep down inside'?," *Psychological Science* 18, no. 3 (2007): 227–229.

4. Paul Ekman, personal communication with Matthieu Ricard, 2003.

5. As the neuropsychiatrist David Galin clearly summarizes, the notion of the "person" is broader, a dynamic continuum extending through time and incorporating various aspects of our corporeal, mental, and social existence. Its boundaries are more fluid. The person can refer to the body ("personal fitness"), intimate thoughts ("a very personal feeling"), character ("a nice person"), social relations ("separating one's personal from one's professional life"), or the human being in general ("respect for one's person"). Its continuity through time allows us to link the representations of ourselves from the past to projections into the future. It denotes how we each differ from each other and reflects our unique qualities. D. Galin, "The Concepts of 'Self,' 'Person,' and 'I,' in Western Psychology and in Buddhism," in *Buddhism & Science, Breaking New Ground*, ed. B. Alan Wallace (New York: Columbia University Press, 2003).

Chapter 5

1. J. D. Haynes and G. Rees, "Predicting the orientation of invisible stimuli from activity in human primary visual cortex," *Nature Neuroscience* 8, no. 5 (2005): 686–691; J. D. Haynes and G. Rees, "Predicting the stream of consciousness from activity in human visual cortex," *Current Biology* 15, no. 14 (2005): 1301–1307; J. D. Haynes, K. Sakai, G. Rees, S. Gilbert, C. Frith, and R. E. Passingham, "Reading hidden intentions in the human brain," *Current Biology* 17, no. 4 (2007): 323–328; W. Singer, "The ongoing search for the neuronal correlate of conscious-

ness." In: T. Metzinger & J. M. Windt (eds.), *Open Mind* (Vol. 36, pp. 1–30) (Frankfurt am Main: Mind Group); W. Singer, "Large scale temporal coordination of cortical activity as a prerequisite for conscious experience." In: *Blackwell Companion to Consciousness*, 2nd edition (in press 2016).

2. K. D. Vohs and J. W. Schooler, "The value of believing in free will: Encouraging a belief in determinism increases cheating," *Psychological Science* 19 (2008): 49–54.

3. Charles Taylor, *Sources of the Self: The Making of Modern Identity* (Cambridge: Harvard University Press, 1989).

4. Francisco J. Varela, *Ethical Know-How: Action, Wisdom, and Cognition* (Stanford: Stanford University Press, 1999).

5. See Lomax's autobiography: E. Lomax, *The Railway Man* (New York: Vintage, 1996).

6. Ibid., p. 276.

7. M. Ricard and T. X Tuan, *The Quantum and the Lotus: A Journey to the Frontiers Where Science and Buddhism Meet* (New York: Broadway Books, 2004).

Chapter 6

1. For a detailed explanation of these possibilities, see M. Bitbol, "Downward causation without foundations," *Synthese* 185 (2012): 233–255.

2. The authors are grateful to Michel Bitbol for directing them to this quote. M. A. Cohen and D. Dennett, "Consciousness cannot be separated from function," *Trends in Cognitive Sciences* 15 (2011): 358–363.

3. Dalai Lama, *The Universe in a Single Atom* (New York: Morgan Road Books, 2005), 122.

4. C. Schnakers and S. Laureys, eds., *Coma and Disorders of Consciousness* (London: Springer, 2012); S. Laureys, *Un si brillant cerveau* (Paris: Odile Jacob, 2015); S. Laureys, A. M. Owen, and N. D. Schiff, "Brain function in coma, vegetative state, and related disorders," *The Lancet Neurology* 3, no. 9 (2004): 537–546; A. M. Owen, M. R. Coleman, M. Boly, M. H. Davis, S. Laureys, and J. D. Pickard, "Detecting awareness in the vegetative state," *Science* 313, no. 5792 (2006): 1402.

5. I. Stevenson, *Twenty Cases Suggestive of Reincarnation*, 2nd ed. (Charlottesville, VA: University of Virginia Press, 1988).

6. See the article written by the three men sent by Gandhi: L. D. Gupta, N. R. Sharma, and T. C., *An Inquiry Into the Case of Shanti Devi* (Delhi: International Aryan League, 1936); Patrice Van Eersel's article in *Clés* 22 (1999), which we have summarized here.

7. P. Van Lommel, R. van Wees, V. Meyers, and I. Elfferich, "Near-death experience in survivors of cardiac arrest: a prospective study in the Netherlands." *The Lancet* 358, no. 9298 (2001): 2039–2045.

8. F. Picard, and A. D. Craig, "Ecstatic epileptic seizures: a potential window on the neural basis for human self-awareness," *Epilepsy & Behavior* 16, no. 3 (2009): 539–546; F. Picard, "State of belief, subjective certainty and bliss as a product of cortical dysfunction," *Cortex* 49, no. 9 (2013): 2494–2500.

9. F. Varela, ed., *Sleeping, Dreaming, and Dying: An Exploration of Consciousness with the Dalai Lama* (Boston: Wisdom Publications, 1997), 216–217.

INDEX